THROUGH ASTRONAUT EYES

PURDUE STUDIES IN AERONAUTICS AND ASTRONAUTICS

James R. Hansen, Series Editor

Purdue Studies in Aeronautics and Astronautics builds on Purdue's leadership in aeronautic and astronautic engineering, as well as the historic accomplishments of many of its luminary alums. Works in the series will explore cutting-edge topics in aeronautics and astronautics enterprises, tell unique stories from the history of flight and space travel, and contemplate the future of human space exploration and colonization.

RECENT BOOKS IN THE SERIES

British Imperial Air Power: The Royal Air Forces and the Defense of Australia and New Zealand Between the World Wars
by Alex M Spencer

A Reluctant Icon: Letters to Neil Armstrong
by James R. Hansen

John Houbolt: The Unsung Hero of the Apollo Moon Landings
by William F. Causey

Dear Neil Armstrong: Letters to the First Man from All Mankind
by James R. Hansen

Piercing the Horizon: The Story of Visionary NASA Chief Tom Paine
by Sunny Tsiao

Calculated Risk: The Supersonic Life and Times of Gus Grissom
by George Leopold

Spacewalker: My Journey in Space and Faith as NASA's Record-Setting Frequent Flyer
by Jerry L. Ross

THROUGH ASTRONAUT EYES

Photographing Early Human Spaceflight

Jennifer K. Levasseur

Purdue University Press
West Lafayette, Indiana

The funding and support of the author by the Smithsonian Institution made the research and writing of this book possible.

Printed in the United States of America.

Cataloging-in-Publication data is on file with the Library of Congress.
Hardback ISBN: 978-1-55753-931-1
epub ISBN: 978-1-55753-933-5
epdf ISBN: 978-1-55753-932-8

Cover images: Astronaut Edward White performing the first U.S. spacewalk, photographed by James McDivitt, June 3, 1965, S65-30431 (NASA); *Earthrise*, photographed by William Anders, December 24, 1968, AS08-14-2383 (NASA); the iconic photograph of Buzz Aldrin, over many years, came to be known by many as the *Moonman* image, AS11-40-5903 (NASA); A whole Earth as seen during the Apollo 17 mission, likely photographed by Harrison Schmitt, AS17-148-22727 (NASA); Harrison Schmitt and the U.S. flag with Earth above, photographed by Eugene Cernan, December 11, 1972, AS17-134-20384 (NASA); Apollo Lunar Camera (Smithsonian National Air and Space Museum).

Smithsonian
National Air and Space Museum

CONTENTS

List of Illustrations

Acknowledgments

I spent my youth admiring astronauts from afar. Southeastern Michigan was a great place for that in the 1980s. John Glenn was the senator on the news from the local Toledo television stations, everyone knew who Neil Armstrong was, and visiting local museums introduced me to names of Michigan-connected astronauts like Edward White, James McDivitt, Alfred Worden, James Irwin, David Scott, and Jack Lousma. Those were astronauts of the past, but they were heroes to me nonetheless. Their mystique was so embedded in my mind that midway through a course on the Italian Renaissance in the fall of 1998 at the University of Michigan, I made up an excuse to leave class so I could go watch John Glenn return to space aboard the Space Shuttle *Discovery* on STS-95. I expected to arrive at the Michigan Union and watch CNN's coverage by myself—I was surely the only person thinking of watching the event. To my surprise, I was late to the party. I became one of around two hundred students with eyes glued to the big screen television, quietly waiting for a specific point in the launch to pass, one fixed in our memories from having watched the loss of *Challenger* live on television as elementary schoolchildren. Nobody moved a muscle until those solid rocket boosters separated successfully. To each one of my companions that day, I say thank you for inspiring my path.

Since the summer of 2002, my professional and intellectual home has been the Space History Department at the Smithsonian National Air and Space Museum. As a freshly minted graduate of George Washington University's American studies program in material culture and museum studies, I took my first steps into collections and exhibitions thanks to two individuals, to whom I owe more than I can express in words for taking a chance on an untested midwestern girl. Toni Thomas and Allan Needell gave me the opportunity of a lifetime as a museum technician, allowing me to prove myself. I worked, as I am wont to do, independently on artifact loans and a project to make artifact information available to the public. Their mentorship resonates in how I continue to approach my work today.

I owe an enormous amount of thanks to the chairs of the Space History Department over the years, who permitted me the flexibility and support needed to work through my PhD program at George Mason University, through the completion of my dissertation, and then this book. Roger Launius, Michael Neufeld, Paul Ceruzzi, Valerie Neal, and Margaret Weitekamp provided guidance and understanding when events like having children came before everything else. In particular, my supervisor, Michael Neufeld, helped shape this book from the time it was still draft dissertation chapters. His editorial skill is unparalleled, and I am thankful for his commitment to my scholarly development. The last steps of preparing this this manuscript were made far easier with the research assistance of Matt Sanders. Likewise, the friendship and encouragement given by other members of my department through the years cannot be overlooked: Martin Collins, James David, David DeVorkin, Cathleen Lewis, Matthew Shindell, and Teasel Muir-Harmony. One benefit of being in this unique scholarly environment is the strength of our fellowship program, which constantly revitalizes those of us long-time residents. To all of those fellows, I thank you for your companionship through our shared journeys, in particular Amy Foster and Matthew Hersch, for their friendship and guidance.

The work of curators hardly happens in isolation, and my most valued moments as a curator occur on projects when the expertise of my colleagues from other departments shines. My sincerest thanks go out to those from the Exhibitions, Collections, Conservation, Archives, Education, and other Museum departments, who make my job easier through their professionalism and expertise. My greatest professional joy comes in our work together, and it is my privilege to team with you to bring amazing stories of space history to our audience. I would be remiss if I did not also thank the leadership of the Smithsonian National Air and Space Museum for providing support along my road to this goal.

Online archives were a staple of my research as they became available over the course of dissertation and then manuscript preparation. Consequently, I made only three in-person visits to archival centers, both in Texas. Somehow, two of those three trips occurred while I was pregnant (with two different children, in 2010 and 2013). I am eternally grateful for the patience of the staff at the University of

Houston Clear Lake's Alfred R. Neumann Library during that 2010 visit, Shelly Kelly and Jean Grant. Jennifer Ross-Nazzal at the NASA Johnson Space Center History Office guided my search for both paper documents and oral histories, and I remain indebted for her continued help and friendship. The finding aid for the NARA selection of NASA records developed by Kent Carter was crucial to navigating a murky selection of Record Group 255 documents, and the staff at the Fort Worth Records Center were patient, kind, and extraordinarily helpful.

This project began as just a simple fascination with a photograph—*Earthrise*, with all of the post hoc symbolism read into the image by the media, environmentalists, and politicians. My first deep consideration of it in the early 2000s when I saw, through their eyes, the moment when three humans first looked upon a revolutionary scene. I wondered what they thought about, what the photographer considered when pointing a camera at that view, and all the moments of discovery that came before and since as seen through somebody else's eyes and cameras. Those three explorers, Frank Borman, James Lovell, and William Anders, acted as our surrogates in that instant, and I need to express my gratitude to them for their courage, insight, and dedication to their mission. To Bill Anders, thank you for opening a window into that sublime moment in your life that inspired the shape of my own.

Not a moment of this experience would mean anything without the support of my family and friends. My parents wanted me to be a scientist or teacher, and they got something in between. I know I made them proud by finding and traveling my own path. Ben, my husband, witnessed the highest of the highs and the lowest of the lows through this journey. There are not words to thank him for such generosity. Our children, Remi and Drew, arrived in the middle of everything but are my guiding lights through the challenges of life. Their laughter, joyous spirits, and love were game changers toward crossing the finish line.

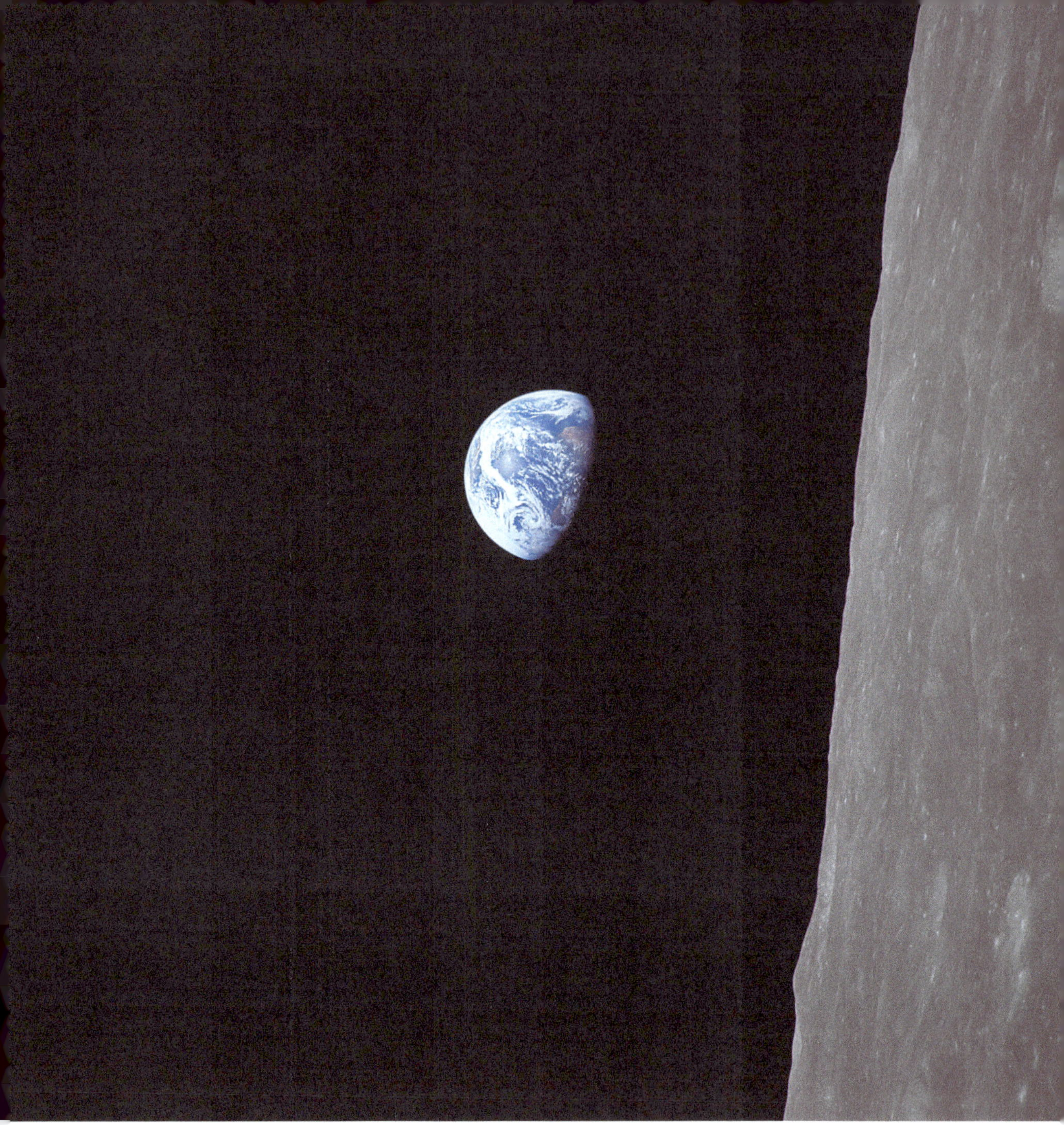

Figure Intro. 1. *Earthrise*, photographed by William Anders, December 24, 1968, AS08-14-2383 (NASA).

INTRODUCTION

Interpreting Astronaut Photography

> *"Through the eyes of the astronauts, we have seen more clearly than ever before this precious earth essence that must be preserved. It might be given a new name borrowed from space language: Earth Shine."*[1]
>
> —*Anne Morrow Lindbergh, 1969*

Two of the most famous Americans in aviation history attended the launch of Apollo 8 on December 21, 1968: Charles and Anne Morrow Lindbergh. Anne later recorded her observations of meeting the astronauts the day before the launch, viewing the Saturn V, and analyzing the implications of what was probably the most meaningful result of the Apollo program: humans seeing Earth from a great distance. In an essay first published in newspapers and then in book form, she reflected on the *Earthrise* photograph that dominated reactions to the mission (Figure Intro. 1). While Charles and Anne shied away from the media intent on documenting the journey of the astronauts, the couple likely felt a kinship with the three Apollo 8 astronauts as visible public figures.

Lindbergh's essay also captured the unspoken and sometimes unexpected benefits of the Apollo program: the spiritual, intellectual, and inspirational aspects of the distant view of Earth. The story of the photograph, a view titled even in mission-planning documents as "Earthrise," symbolized the culmination of almost a decade of astronaut photographic activity at NASA. *Earthrise* became one of the defining visual moments of Apollo, combining the technical and mechanical with the ethereal and emotional. The image expresses, in one frame, the sublimity of engineering expertise in a politically driven nation. *Earthrise* became the single best visual expression of traditional notions of American exceptionalism during the so-called long 1960s.[2] Americans traveled to the Moon and returned having "conquered" the space "frontier."

Astronaut photography has profound importance to the fabric of American culture, with the images shaping shared ideals about American life and society for decades after. How we remember those events, either through living or shared cultural memories, depends greatly on photographs taken during key moments. I argue that more than print or television coverage, astronaut photographs shaped our collective memory of a period punctuated by moving and still images of assassinations, urban riots, and military activity. Yet rarely do scholarly works even consider astronaut photographs as part of the visual fabric of the long 1960s. These images are more than just photographs to wow the viewer—embodied in the images we most frequently encounter are the political, cultural, and social symbols of the Space Race. This book seeks to redirect attention to this period with a wider lens, integrating astronaut-captured photographs into the visual narrative of the period and assessing the role astronauts played in shaping public memory. Cameras were the tools to tell the story of spaceflight. The images created imprinted lasting visual memories on our shared cultural consciousness.

My examination of astronaut photography utilizes visuals as a means of understanding both their influence on, and place in, a complicated visual landscape. Popular and historical scholarship regularly characterizes this period as the height of visual chaos with television news, magazines, art, and other media presenting a battery of images to confront and sometimes offend the viewer's senses. In complete contrast, NASA provided images of humans in space, technological achievements, and the discoveries of space exploration. Those inside and outside the space program consistently used powerful visuals in addition to well-worn rhetoric to promote space exploration as a core mission, affirming President Kennedy's vision of American culture. Using photographs as tools to influence policymakers and the public alike reflected the interpretive nature of photography. As a tool for public relations, engineering documentation, scientific investigation, and recording the experience itself, astronaut photography opened the door to understanding human space exploration.

To contextualize the images historically requires a framework for interpreting them. Scholarship in four areas informs my argument and shapes the organization of this book: technology, communications, visual culture, and public memory. These topics have well-established historiographies, but very few instances, even unrelated to space history, attempt to examine astronaut photography using these frameworks. My study extends recent scholarship that sought to engage with similar materials or subject matter, but this interpretation is unique in its focus on astronaut images seen in terms of the visual culture of the 1960s.[3] I demonstrate how deeply these images penetrated American culture because of their broad intent, availability, circulation, and appeal to audiences, far and beyond that of exploration of the past.[4] These visual products, created by our proxies in space, became a significant point of departure for American culture during the Cold War, symbolizing a psychological victory despite the tragedies of the period.

Some boundaries are in order for any historical work. I tell this story from a specifically American perspective because the images were provided by American astronauts for a predominantly American audience. Their effects and ultimate legacy are, however, clearly global. I also bound this book by the years dedicated to accomplishing the most audacious American goal of the Cold War: sending a human to the Moon. This early period, sometimes referred to as the heroic age of human spaceflight, began with NASA's selection of astronauts in 1959 and the end of Project Apollo in December 1972. Excluded from this study is photography beyond Apollo. By the time of Skylab in 1973 and the later Space Shuttle program of the 1980s, changes inside and outside NASA shifted the focus of the human spaceflight program. Camera types expanded, scientific uses grew dramatically, and the results changed as the astronauts do not leave Earth orbit, a subject worthy of a separate study of its own.

For historians, research material often takes the form of written words in sources such as diaries, memos, letters, and other textual material. The wordless nature of images makes them less comfortable territory. Reading an image for information, as art historians and visual culture experts do, is an acquired skill, and the potential for a multiplicity of interpretations can frustrate the best historians. In the case of astronaut photography, scholars from a range of backgrounds have engaged with the images as both evidence and art. I seek to forge a path through the middle, engaging with the material in terms of the visual constructed by the written.

Unpacking this subject using a single historiographical context such as visual culture, art history, technology, or public memory too narrowly defines the study, although all four used equally would not take into account the nature of resources and documentation available. Some technical and nontechnical studies published since the Apollo program engaged with astronaut photographs on some level, but typically only as illustrations of a point about a larger part of the program. My study requires the interpretation of documentary, photographic, and technological evidence to understand the cultural significance of astronaut photography. By incorporating a wide set of materials in this study, including a close visual reading of key photographs, I aim to move beyond intellectual monographs or narrow technological histories that make little use of the rich primary sources and oral histories.[5] My goal is to not only address the specifics of mission photography that lead to iconic astronaut photographs, but also examine them within the context of photography, the power of images, and their ongoing cultural resonance.

EXPLORATION AND TECHNOLOGY

Within the history of exploration, camera technology offered the opportunity to chronicle journeys for science, commercialization, and public relations. The concept of exploration implies a harsh working environment, with potential pitfalls and situations explorers cannot predict. Photographic work in such conditions presents additional challenges, whether the conditions are the diverse terrain of the unexplored American West in the late nineteenth century, the treacherous Antarctic ice sheet around the turn of the twentieth century, or the cold vacuum of space.

This study is about not only early astronaut photography, but also the related historical situations of imaging in extreme conditions. Recent scholarship from Elizabeth Kessler illustrated a formula similar to the one I use here by connecting visual culture and technological capabilities in exploration settings. Kessler's study interprets Hubble Space Telescope images as the byproduct of both scientific data and an existing understanding of and connection with landscape painting from the nineteenth century.[6] My work likewise examines the process: technological choices made along the way and training test pilots to do work traditionally done by professional photographers. Prior expeditionary projects, those to the American West and Antarctic regions, employed trained photographers, men who by education or trade knew the technology and how to maximize issues of lighting, exposure, and composition. In the case of astronaut photography, its position within mission-planning priorities was far from the top. While professional photographers employed by NASA assisted in selecting camera equipment, training the astronauts in camera use, and developing the films postflight, the art of making a well-composed and elegant photograph was not intentional or sometimes possible for astronaut-photographers. Time and location usually did not allow for much flexibility. None of the hallmarks of what define a professional artistic photographer, other than capturing images with a camera, was present in the experiences of the astronauts responsible for memorable views of our first forays into outer space.

Astronaut photography also fits within the discipline of the history of technology. Specialized histories of spaceflight technologies abound, including a few on cameras specifically, but none of those approaches the serious contemporary scholarship that shapes the field.[7] Of particular interest are works by Philip Scranton on a community-based approach to understanding technological developments in spaceflight and their influence within society, and David Mindell on the human-machine interface as seen with the development of the Apollo computers.[8] These multidisciplinary approaches to the history of technology delve into the more complex stories behind the creation of technologies as opposed to the focused, fetishized histories written previously, histories that focused on what Scranton calls the "mastery of technology." This study, therefore, takes a multidisciplinary approach to avoid the well-trodden ground of specialized camera technology histories so as not to prize the detail over an assessment of its role in changing American cultural memory. My explanation of astronaut photography, including the cameras selected and used during flight, offers a unique opportunity to see, in the literal sense, a type of technology as a fundamental part of how American culture and memory continue to represent and understand the spaceflight experience. Critical to my formation of that argument is Ruth Schwartz Cowan's landmark piece on the social construction of technologies through networks of consumer decisions.[9] Her analysis would suggest, in the case of my study, that the position of consumer inside a network could be filled by those at NASA selecting cameras and in audiences outside of NASA using the images.

Coupled with these nuanced examinations of spaceflight history, broad technological histories help shape the direction of this book. David Nye's *American Technological Sublime* offers a well-reasoned framework for how the sight of technological achievements shaped the American understanding of them, while calling attention to similarities between views of the American West and spaceflight.[10] My argument, in concert with Nye's characterization of how such experiences can bind a society together, is that this process was made possible in the Space Age through the images captured by the astronauts. Anyone could see what astronauts achieved. We now live in a society that expects amazing technological achievements as the norm, and we require visuals to believe them. In the fast-paced, Twitter-ized culture of the early twenty-first century, one might ask if even photographs, so often scrutinized today for signs of surreptitious Photoshopping, will remain the gold standard as evidence of the spectacles of space exploration.

COMMUNICATING SPACEFLIGHT

What makes astronaut photography such an interesting topic is both its deep penetration into American culture and its global spread as a means of communicating the accomplishments of NASA engineers, scientists, and astronauts. There was an implicit trust in the mimetic quality of the images. Roland Barthes explained that some visual evidence requires no language or key because "for all the kinds of images only the photograph is able to transmit the (literal) information without forming it by means of discontinuous signs and rules of transformation."[11] While the audience for astronaut photography may be broad, the astronaut-photographers were by no means artists. It was the images, far more than their words, that became the literal focal point for most memories of the early human spaceflight program.

By examining astronaut photography, I am seeking to expand our historical and cultural understanding of the early space program in a specifically American context. Moving and still images add richness, texture, and dimension to life otherwise captured in impressions made through words on paper or in memories saved in our minds. Sigmund Freud called a photograph the screen onto which we project visual memories.[12] Similarly, theorist Siegfried Kracauer posited that photography is an attempt to capture a spatial continuum while history captures the temporal continuum.[13] The generation of a spatial concept of a historical moment, one passed but not necessarily experienced personally, required accurate photographs and information, especially for a public already immersed in the visual documentation of contemporary life. Seeing events from around the world on the nightly news or in the pages of daily newspapers and weekly magazines linked the possibility of remembering global events through associated visuals, not just in the words generated by journalists.

Because a wide audience viewed the images, the range of possible reactions was equally broad. Those who financially supported expeditions may have found satisfaction in the economic development possibilities seen in images: potential mineral deposits or natural resources seen in photographs of western American regions, or the lack of resources in polar images. Michel Foucault addressed this in his discussion of power, explaining that one expression of power is the control of images. In the case of astronaut photography, both NASA and the astronauts wielded power as the arbiters of image content.[14] Institutions define many of the rules by which these elements of the cultural communication occur, but the interpretation of images remains in the minds of the audiences. Their needs shaped the ways in which the photography was distributed. In each case examined here, however, the broadest version of the public audience was not the intended primary consumer of the photographs, but rather those who chose intentionally to employ them for their own purposes. With so many audiences, interesting and dynamic possibilities for reaction and interpretation existed.

Using Barthes' theory of images, my study also examines the symbolic and literal messages (connoted and denoted, in his terms) of astronaut photography. Reading these images brings forth a variety of impressions, associations, and memories. Literal messages in spaceflight are obvious: a rock was in a specific position when collected by an astronaut. Denoted messages convey information in similar literal ways to those seen in scientific or medical photographs. Connoted messages vary by the subjective attitudes, beliefs, and other psychological filters used by people when analyzing visual information. To read an image, to process information from it in a strategic way, requires a level of interpretation that brings forward elements seen as more dominant or important and pushes to the side elements deemed less significant.[15] This book explores the production of these symbolic and literal messages within the historical, cultural, and political contexts of this early period of human spaceflight.

What is essentially a visual culture study set within the context of examinations of human spaceflight requires looking outside of space history literature to secondary works that guide it toward a place within other historiographical narratives. Studies abound of the meaning of photography of newly explored places and the function of photography in culture.[16] To focus my work, I call attention to what is essential to the narrative of astronaut photographic reception, situating all of these issues within a framework established by other histories of exploration photography. At its core, the story of astronaut photography is about the imaging of exploration by humans, but not necessarily professionals; a visual record created by the few who made the journey for those who did not. Across the scope of exploration projects over the last century and a half, these experiences stand as key moments in which ideas of success, failure, pride, and discovery depended on a visual record.

VISUAL CULTURE AND SPACEFLIGHT

Visual culture, the study of the relationship between the visual and the consumer, provides a framework with which to analyze the production, dissemination, and reaction to astronaut photography. Leveraging this scholarship helps determine what the astronaut photographs mean in American culture, from the turbulent years during which the United States went to the Moon through to the present. Approaching them this way also opens up opportunities to see where astronaut photography shares a visual legacy with works from the history of photography and exploration. Exploration with cameras in the century prior to Apollo cameras created a visual catalog rich in rhetoric that informed, though only subconsciously, the types of images NASA sought to distribute to viewers. To study the visual culture of space exploration means, as other scholars have done with exploration of the American West and Antarctica, examining how the images were made, circulated, and understood by audiences. That approach shapes why I assert that these images are critical to understanding the period in which they were captured and what they continue to mean to viewers today.

American notions of the frontier and national ideals guided early perceptions of astronaut photography, but readings of the images challenge the use of such rhetoric in relationship to space exploration. NASA refrained from assigning grand meanings to astronaut photographs in the text accompanying images, choosing to supply only basic information in captions and press releases. Notably, social commentators, including Archibald MacLeish and James Dickey, wrote essays about the implications of astronaut photography.[17] Through images, people could feel as though they participated in a historic moment that only a handful of

Americans actually experienced: circling the Earth or Moon in spacecraft barely larger than midsize sedans.[18] As Daniel Boorstin stated at the beginning of the Space Age, "[Travelers] found amazement and delight and have reflected that life back home need not always remain what it has been . . . there are more things in heaven and earth than was dreamt of in their philosophy, that the possibilities of life are not exhausted on Main Street."[19] The public were to some degree guided in their understanding of the images by noted writers, coming to understand astronaut photography in terms other than those NASA expected. Astronaut voyages in space provided the sense of the sublime sought by audiences, providing an escape from daily reality.

Within the visual evidence produced during the space program lay three general media types available for examination: movies, still images, and art. To maintain consistent and focused terminology, technology, and evidence, selecting one medium is necessary. Scholars Kathy Keltner-Previs and Ann Goodyear addressed visual communication during the early space program, but they focused on television and art, respectively, and not the production, dissemination, and reception of still images to form a complete visual narrative.[20] Still photographs distributed broadly via local, national, and global news outlets in printed form make for a sufficiently narrow but interesting study. Television and movie footage, in the form of broadcast signals and film, provided media for production teams to display space accomplishments for a fleeting moment. Until fairly recently with digital technology and the Internet, moving images were far more difficult to reproduce and share, resulting in a less permanent place in public memory.

A critical method for conveying space successes, my study contends, are the photographs created by participants in the experiences of leaving Earth. This may be associated with the overlap between early human spaceflight and W. J. T. Mitchell's definition of a period he calls the *pictorial turn*: when American culture moved from reliance on the printed word to visual media for news information.[21] Images astronauts returned from space became vital contributions to public memory of human spaceflight as well as critical pieces of evidence for scientific research. As visual scholar John Tagg points out, however, the camera is never neutral.[22] The U.S. government and particularly NASA seemed to sway public opinion with stunning photographs taken by astronauts. But in reality, even images could not push support for space exploration much beyond the 50 percent mark at the height of NASA's popularity in the late 1960s.[23] Apparently, even the seeming unimpeachable status of photographs did not drive public opinion or encourage additional political support for funding human spaceflight once NASA achieved Kennedy's lunar goal. Visual, scientific, and technological benefits proved too little in comparison to the social, economic, and political pressures of the 1970s.

REMEMBERING OUR HEROES

Historian David Lubin made the case for how iconic images become entrenched in public memory because they revisit previously seen images with familiar content.[24] The ease of obtaining printed reproductions of photographs in newspapers or magazines throughout the last half of the twentieth century meant they often served as commemorative items of historic moments. From the perspective of at least one person, Richard Underwood, who trained early American astronauts in photographic techniques, the quality of astronaut photographs was the key to their "immortality" as explorers.[25] As one present in the moment, it was easy to overemphasize the potential of photographs, but Underwood's suggestion considered the possible disconnect between astronauts and images that could unfold long after the days of the heroic exploration of space. As Benjamin Lazier points out, "It is one thing to trace the spread of use of the images themselves . . . but it is something else to track how the planetary horizons afforded by photographs of the whole Earth have surmounted, inflected, complemented, or corrupted the earth-bound horizons of everyday experience."[26] Still photography, however, may well be the best, most important source of fixing permanently in public memory the success of the U.S. space program of the 1960s.

While many people remember astronaut achievements before the contributions of hundreds of thousands of people who enabled their flights, I see a disconnection of astronauts themselves from public memory because of the management of the photography. Rarely do astronauts appear in mission photographs, and often those seen outside the spacecraft are for purposes of scale, not really posed or planned. Some exceptions to this observation are the images recognizable as icons of the period: Ed White spacewalking with the backdrop of the Earth (Figure Intro. 2) or Buzz Aldrin saluting the U.S. flag on the Moon (Figure Intro. 3). With limited examples of images that serve as evidence of human activity in space, I argue a general disassociation exists between specific personal information in images and

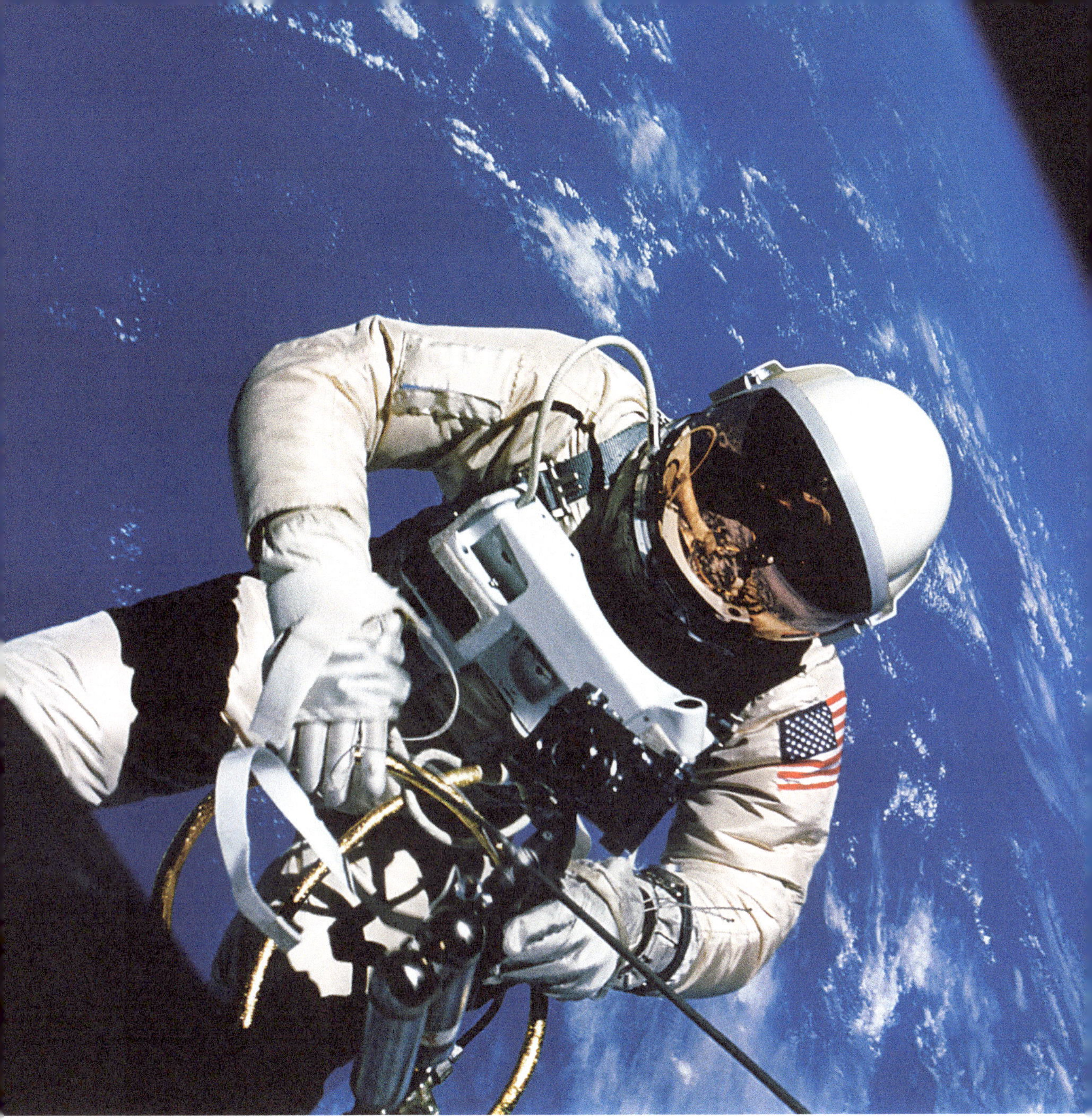

Figure Intro. 2. Astronaut Edward White performing the first U.S. spacewalk, photographed by James McDivitt, June 3, 1965, S65-30431 (NASA).

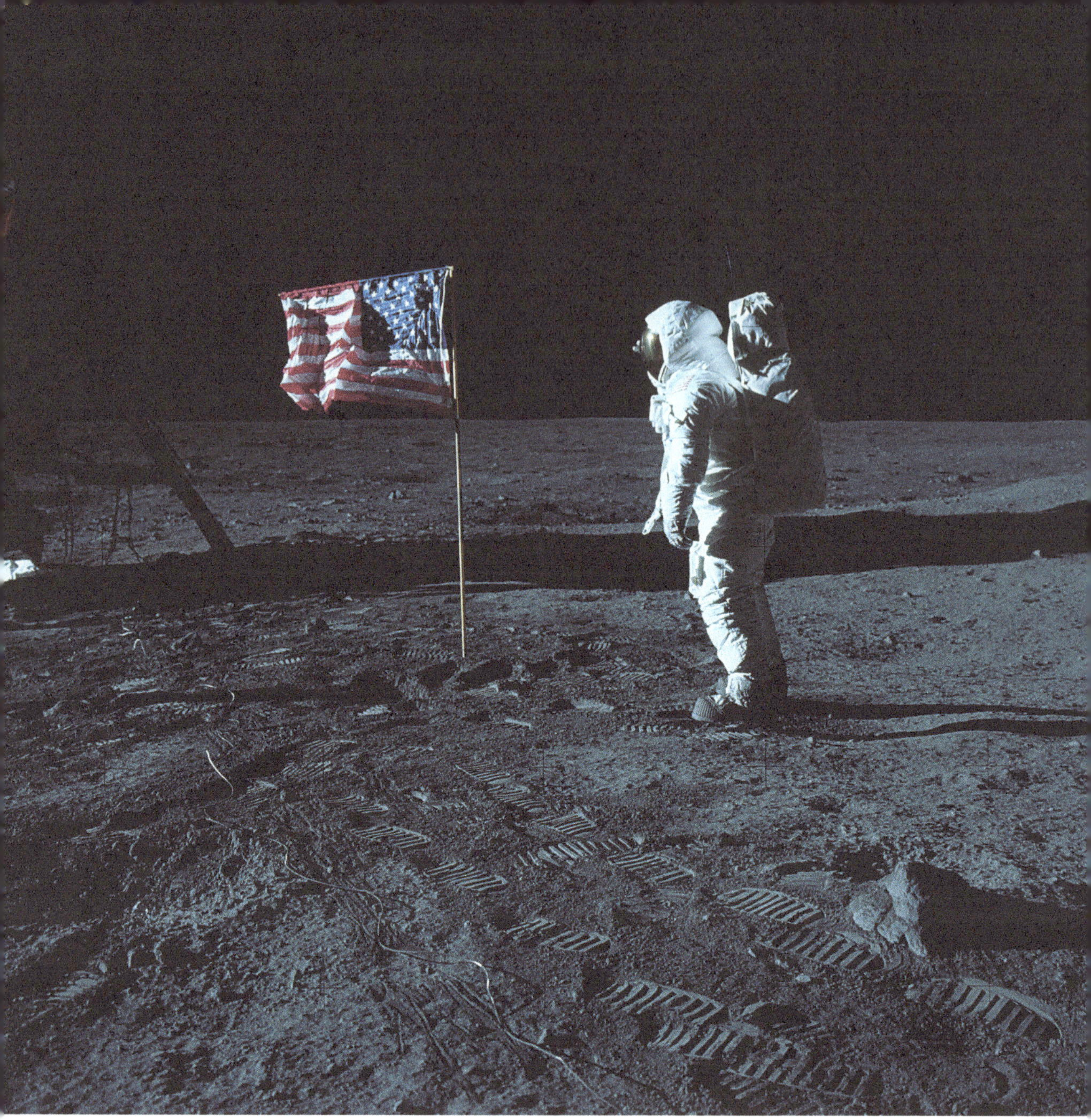

Figure Intro. 3. Astronaut Buzz Aldrin saluting the U.S. flag on the Moon during the Apollo 11 mission, July 20, 1969, AS11-40-5874 (NASA).

the generalized impressions that endure in public memory. Many people realize a vague fact such as "humans did it," but not that Gene Cernan was commander of Apollo 17, the last human mission to the Moon. Underwood's statement about immortality then applies in a general sense to the images and activities of astronauts, not necessarily the exact people who carried them out.

While American historical scholarship tends to neglect spaceflight and astronaut photography as a serious topic, ongoing usage and consideration of the images reflects their cemented place in public memory. The formation of those memories, as argued by Maurice Halbwachs and memory scholars since involves the mixing of individual and socially constructed memories, one informing the other either intentionally or not.[27] Fundamental to perpetuating the memory of early human spaceflight, memory scholars would suggest, are the constant reminders about the events. In the case of the first person walking on the Moon, NASA uses the media and especially online social media tools to provide consistent and constant visual reflections. These spaces for focusing shared memories become what Pierre Nora called *lieux de mémoire*, sites of memory found at intersections between history and memory.[28] In the context of this book, these sites, more frequently than not, feature astronaut photography.[29] Collective remembrance scholarship since Halbwachs even engages with how when individual memories do not exist (for example, in the case of those not alive when Armstrong's first steps took place), ongoing commemoration of the past can serve to reinforce collective memory.[30] I argue that NASA astronaut images serve as what scholar Alison Landsberg refers to as prosthetic memories, which allow personal connections to moments of the past because of the emergence of mass culture, even when those events were not experienced personally.[31] The ease of duplicating images, then as now, means that the perpetual publication of astronaut images in commemoration of early human spaceflight keeps the memory of that period alive and well, especially during significant anniversaries. The photographs inscribed upon NASA and the astronauts an emotional and indelible link to the public memory Richard Underwood predicted in his preparation of astronauts for their photographic work.

Photography by astronauts throughout the space program, to include the Space Shuttle and International Space Station programs, have been devices for stimulating and reviving positive feelings about the successes of human spaceflight. Companies across the United States and some in Europe who contributed technological elements to NASA's programs have capitalized on that success. Through visually compelling advertisements and press kits, those companies used publicly accessible astronaut photographs to remind viewers of their role in a visually compelling way.[32] Such connections linger even today with the marking of Omega watches, Hasselblad cameras, General Motors, and Plantronics using effective visual references to their involvement in the Apollo program.[33] For print media especially, the availability of a wealth of stunning imagery meant a nearly endless ability to access the good will that still connects public memory with the early space program of the United States.

Despite astronaut photography serving to connect current public memory with early human spaceflight, our cultural literacy about this period is open to question. Astronaut images are forms of mass media that contain unrealized potential to transcend time and shape cultural memories. As those triumphant moments move farther into the past and generations with no personal memories of the Cold War Space Race make decisions about the shape of space exploration, the power of astronaut photography as a means of perpetuating public memory should mean a greater attention to why they have such significance. Their meaning, if transcendent, will survive and thrive long past the lives of anyone involved in making the photographs a reality, as they nearly have already.

Evaluating the place of astronaut photography within our shared American cultural memory requires stepping back to look at the creation of those images as part of the relationship between producers and their audiences. We often accept as granted a certain ease of capturing photographs using modern technology at our fingertips, but astronauts, launching atop massive rockets and taking risks to simply survive in space, spent what time they could sharing the look of space through photos and the feel of space through their words. Allowing astronauts to step back from spacecraft operations and secure visual representations of what they saw required NASA to go about finding the right equipment, process, and distribution to solidify those images in the public consciousness.

CHAPTER 1

Why an Amateur Needs a Better Camera than a Professional

> *"A pro can do professional work because he is familiar with the principles of light and optics. Few amateurs have the time to master this science so thoroughly. There is a camera, however, the single lens reflex Hasselblad, that will put a more professional quality into any amateur's work, even a beginner's. It does through precision, quality lenses and automatic aids."*[1]
>
> —*Hasselblad 500C print ad, July 1961*

In June 1965, astronaut trainer Richard Underwood guided Robert Gilruth and other NASA managers to the Building 8 photo lab at the Manned Spacecraft Center (MSC) in Houston to review Gemini IV mission photography. Underwood, a professional photographer by training, knew instinctually that what the group was about to see could shift perceptions of Earth and human space exploration. The group reviewed photographs from the Zeiss Contarex camera used during Edward White's spacewalk (EVA, or extravehicular activity) and from the Hasselblad camera employed by mission commander James McDivitt. Gilruth (then MSC director) and Underwood debated the value of the images. Their review focused first on images of the spacecraft and astronauts, but Underwood intentionally refocused attention on images of Earth. Underwood attempted to convince Gilruth of the scientific and public relations value of those images. His characterization of the photography suggested ways in which astronaut photography could support the NASA mission, giving substance to the rhetoric and bolstering financial support from the government: "[W]e're looking at things that no human being had ever seen before, parts of Africa and other places. You can see what's really going on."[2] To Underwood, more cameras and better training would carry an important visual message to audiences: real people were seeing Earth from space, offering new perspectives on our globe and ourselves. Impressed by the photographs before them, Gilruth, George Low (deputy center director), and Maxime Faget (spacecraft designer) agreed with Underwood and ordered an increase to resources for astronaut photography.

Midway through the 1960s, NASA's Project Gemini flights provided engineers, managers, and astronauts the opportunity to develop and practice procedures deemed critical to lunar landing missions. On just the second flight, Gemini IV in June 1965, White became the second human to venture outside a spacecraft and into the vacuum of space, a moment McDivitt captured on 70mm Kodak film using a professional-quality camera modified for the unique needs of an astronaut.

The spacewalk and Earth views McDivitt captured from the relative safety of the cabin did more than provide a moment of pride for NASA and a nation invested in the success of its human space program. Handheld astronaut photography had been a very small part of the four orbital Mercury flights in 1962 and 1963, and the first flight of Gemini in 1965, so it was not until the review of Gemini IV photography that motivations, rhetoric, and physical demands turned astronauts into photographers in the spirit of earlier expeditions. Having a good camera that functioned in the space environment was essential to giving these nonprofessional photographers an opportunity to return quality images of the extraordinary views outside our atmosphere.

The new perspective shown in Gemini IV images turned eyes down on Earth, the sublimity prompting NASA managers to support astronauts with the right tools. Because of their post-flight experience in the photo lab, Gilruth charged Underwood with instructing astronauts not only in the technical aspects of photography, but also in photography as a way of thinking about audiences. Reviewing Gemini IV photography, seeing both the human in space and the Earth from space, encouraged John Brinkmann, head of NASA's Photographic Technology Lab (PTL), Underwood, and the rest of those involved in photography work toward spectacular images to appeal to the public and provide information for scientific, technical, and political ends.[3] A human against the backdrop of space gave viewers a sense of scale while also exciting them about the potential of human exploration. Gilruth, in his role as head of the facility responsible for building human-piloted spacecraft, astronaut training, and mission operations, was certainly aware of what those photographs could do for NASA.

While Project Mercury and Gemini engineers and managers apparently never questioned the value of photographing human missions for engineering documentation, they had a harder time understanding the point of photography, filmmaking, or television coverage for nontechnical purposes. Historian Roger Launius notes that putting U.S. astronauts in space had little to do with understanding the environment, doing scientific experiments, or anything other than surpassing Soviet achievements and winning the Cold War Space Race.[4] Matthew Hersch highlights the tension between the first groups of astronauts and the scientific and technical community at NASA, and shows how as a group, the astronauts could enact their own practices or resist changes to mission planning based on their role as the face of the program.[5] Still photography was mostly an outgrowth of astronaut experience, interactions with professional photographers, and knowledge of a public hungry for images of the astronauts' lives and experiences as the country's first space voyagers.

Researching the origins of U.S. astronaut photography presents a serious challenge to scholars today. Not only have many of the astronaut-photographers passed away (including all of the Mercury 7), but so have many of the engineers and technicians employed by NASA and the camera manufacturers. The paper trail of their work appears lost for the most part. The Johnson Space Center's records at the National Archives center in Fort Worth, Texas, include no materials from the PTL, where films returned from flight were processed. Whether those records fell under federal record regulations remains unclear. The difficulty locating sources becomes perhaps the greatest challenge to interpreting the story of astronaut photography. By balancing information gleaned from oral histories against the modest number of primary sources, such as mission reports, correspondence, and technical information maintained by the Flight Crew Operations Division, one can form a reasonable picture of the way in which NASA selected cameras for spaceflight and for what ends.

The cameras themselves also complicate research, as they reside in multiple locations, such as the Smithsonian National Air and Space Museum, the Cosmosphere in Kansas, the Visitor Center at the Kennedy Space Center in Florida, and the Johnson Space Center in Texas and its visitor center, Space Center Houston. While

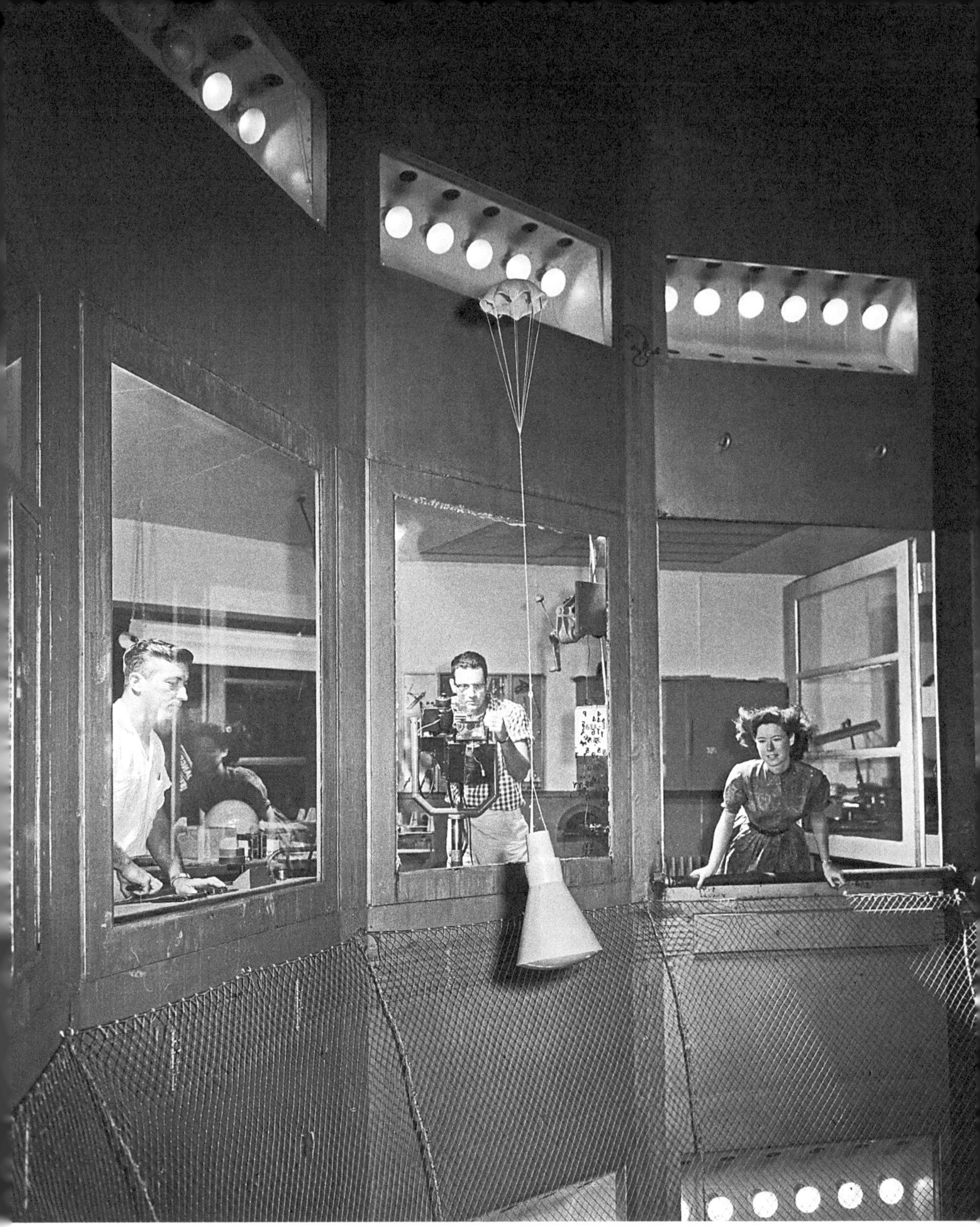

Figure 1.1. Mercury capsule model in spin tunnel at the Langley Research Center, September 11, 1959, LARC Image #L-1959-06212 (NASA).

the first two museums maintain formal accession documentation and transfer paperwork from NASA, the provenance of items in the hands of the NASA centers and their associated visitor centers is less clear. Through the years, documents and metadata often omitted recognition of their status as space-flown artifacts. Locating evidence of their identity is problematic at best, and usually impossible. A long-term goal in my capacity as the responsible curator for astronaut cameras in the National Collection of the Smithsonian has been to understand the disposition of these cameras generally in order to preserve the material legacy of technological choices made for the space program, which supplied a rich visual record of these dramatic moments in human history. Together, the material culture, oral histories, and primary sources enrich our understanding of the importance of astronaut photography to the visual culture of this period of the Cold War.

HANDHELD PHOTOGRAPHY IN PROJECT MERCURY

Engineers, going back to the space agency's prior incarnation as the National Advisory Committee for Aeronautics (NACA), needed photographers to document experiments for aircraft designs, rocket launches, and other tests around the country. Work at the Ames Research Center in Mountain View, California, the Lewis (now Glenn) Research Center in Cleveland, Ohio, the Langley Research Center in Hampton, Virginia, and the Wallops Flight Facility on Wallops Island, Virginia, necessitated thorough visual recording for later analysis. NACA sought out photographers, many with aerial reconnaissance photography experience in World War II, to depict this work from the ground and the air. Brinkmann, NACA's lead photographer during this period, explained in his official NASA oral history how a group of engineers and photographers would take boats out to Wallops or nearby islands to watch launches directed by Dr. Robert Gilruth, one of NACA's most senior engineers.[6] Other tests photographed included wind tunnel work that employed elaborate photographic systems to show flow rates over objects and burn processes on ablative materials, usually done with fast-motion photography to allow engineers to slow down the frames and see what was occurring (Figure 1.1).

This cadre of photographers documenting NACA's early experimentation with aircraft and spacecraft design transitioned to the new space agency created by President Eisenhower in 1958. Owing to their work at Langley, Brinkmann's division was absorbed into the Space Task Group (STG) headed by Gilruth, and assigned the duty of documenting STG's human spaceflight project, Mercury, which aimed to put a person into orbit. The photographers had just as many new experiences to grapple with as the engineers. Eugene Edmonds, an early hire by Brinkmann, recalled in his official oral history: "Well, they'd want everything photographed, so I started out in a helicopter, following these planes up about a thousand feet . . . it reached the point where they were going to start firing it off of a rocket up at Wallops Island, Virginia, and they were going to fire it, say, about ten thousand feet, and they wanted that covered."[7]

NASA staff photographers of the early 1960s, like those in other professional photography roles, knew the landscape of available technologies needed to do their jobs, which in turn informed their recommendations for astronaut use. At the time, simple point-and-shoot photography abounded for professionals as well as the public thanks to the wide availability of 35mm cameras, instant cameras like Polaroids, and single-lens and twin-lens reflex cameras to produce different image formats from a single experiment or session. Options from companies in Germany, Sweden, Japan, Korea, and elsewhere competed for market share, and like any other professional photographers, NASA staff photographers had preferences for what equipment they carried based on the needs of the image users.[8] And frequently, for engineering tests, staff required specialized movie cameras that captured images at high rates in order to examine details of tests at slow replay speeds.[9]

The STG's sole purpose, though, was putting Americans in space to best the Soviet Union.[10] After President Kennedy expanded NASA's human spaceflight goals in May 1961 to include a lunar landing, experienced photographers like Brinkmann, Edmonds, and John Holland picked up shop and moved in 1961 and 1962 to Houston, Texas. As part of the new Manned Spacecraft Center, the photographers and their colleagues worked in buildings around town while awaiting the construction of a campus near Clear Lake (after 1973 called the Lyndon B. Johnson Space Center).[11] Simultaneous to the photographic team's work, NASA assembled a group of seven test pilots to fulfill the goals of the STG.

In April 1959, NASA announced that talented jet test pilots from the United States military services were selected to become the first American astronauts. Their names and faces became synonymous with the archetypal American hero: Alan Shepard, Virgil "Gus" Grissom, John Glenn, Scott Carpenter, Walter Schirra, Gordon Cooper, and Donald "Deke" Slayton. Redstone and Atlas military missiles, refashioned as launch vehicles with single-passenger Mercury capsules atop, carried six of these national icons to first the edge of space and then orbit. Shepard and Grissom made short, fifteen-minute flights in 1961 that gave them a taste of the spaceflight experience, while Glenn and Carpenter (who replaced a medically disqualified Slayton) followed the next year with three-orbit missions. Schirra's late 1962 flight and Cooper's in early 1963 wrapped up the program with extended stays in orbit, laying the groundwork for performing more experiments in space over longer missions that marked Mercury's successor program, Project Gemini.

In the eyes of many Americans, these seven could do no wrong. Photographs of their silver spacesuited frames graced the covers of newspapers and magazines around the world, with their wives and families gaining notoriety as the foundations of these idealized American lives. NASA promoted the Mercury 7 endlessly, making these men larger than life, the faces of a program that employed hundreds of thousands to put them into space. Of course, that pristine exterior belied the normal, complicated lives they led—just outside the glare of the spotlight.

With the move to Houston and increased activities at the Cape Canaveral launch facilities, photographers under Brinkmann's charge rarely wanted for work. Between establishing an operational processing laboratory in old airplane hangars at Ellington Air Force Base (near the future site of MSC) and photographing the testing of astronauts at training facilities around the country, the Photographic Division also played a role in suggesting cameras for the spacecraft. Brinkmann described the managerial and equipment set-up process as the easy part, but putting cameras in the spacecraft was a bit of an unknown.[12] Integrating photography into the spacecraft would result in a new set of rules, restrictions, and demands. As a professional photographer, he also saw the possibility of conflict between the needs of NASA engineers (concerned with safety, weight, and tests) and the influence visual media could have in making spaceflight real for people back on Earth.

Documentation from this period is sadly lacking, as small equipment was acquired on an ad hoc basis, unlike the large engineering projects to build infrastructure to test rocket engines and launch spacecraft. A small camera recorded Alan Shepard's Mercury flight in May 1961, the first U.S. human spaceflight, using a mirror adapter pointing out a small capsule porthole. The J. A. Maurer-manufactured Earth/Sky Observer captured a series of images at the rate of one frame per six seconds, giving an almost movie-like quality when the entire set of images plays in sequence for the fifteen-minute flight (Figure 1.2). Whether selected by the spacecraft manufacturer (McDonnell Aircraft Corporation of St. Louis, Missouri, for the Mercury and Gemini capsules) or engineers at NASA, small 16mm movie cameras made their way into the Mercury spacecraft, embedded pieces of equipment that kept them out of the astronauts' way and presented wide viewing angles for detailed post-flight analysis. The Pilot Observer and Instrument Observer cameras, manufactured by D. B. Milliken Company of Arcadia, California, recorded just what their names indicate (Figure 1.3). These films showed the pilot's activities directly and in relation to the instrument panel. As a backup starting with Gus Grissom's *Liberty Bell 7* flight in July 1961, astronauts wore a circular mirror on their chests so that the Pilot Observer Camera would also record the reflection of the pilot's use of the instrument panel, just in case the Instrument Observer Camera malfunctioned (Figure 1.4).

While these sequence cameras operated throughout the Project Mercury flights and early Gemini two-astronaut missions, smaller handheld 16mm movie cameras took their place during Gemini and Apollo in order to make the documentation tool more mobile. It was mobility, longer missions, and the need for visual documentary representations of what people saw from the window of the spacecraft that brought about the era of handheld astronaut photography.

The astronauts played a role in the development of the tools they would use, including the spacecraft, spacesuits, command and control systems, and other equipment necessary to their work in space. NASA engineers demonstrated a willingness

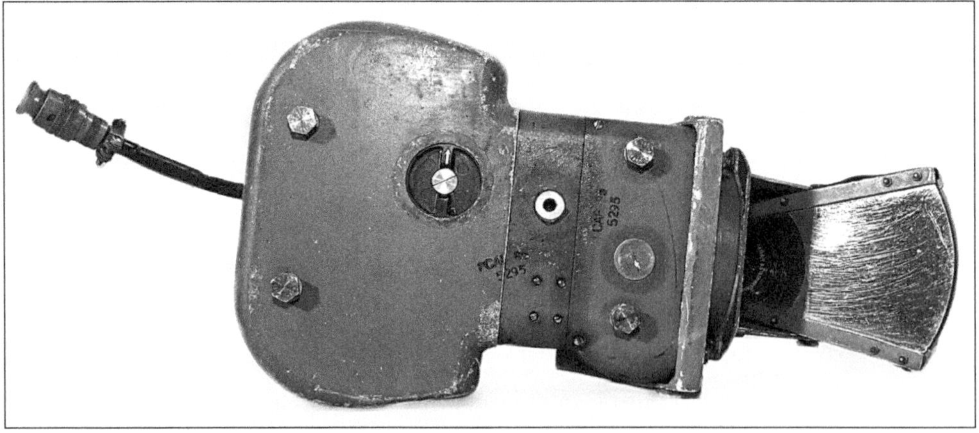

Figure 1.2. Earth/Sky Observer camera manufactured by J. A. Maurer, used on unpiloted Mercury-Atlas 3A mission on April 25, 1961, NASM Cat. #A19781516000 (Smithsonian National Air and Space Museum).

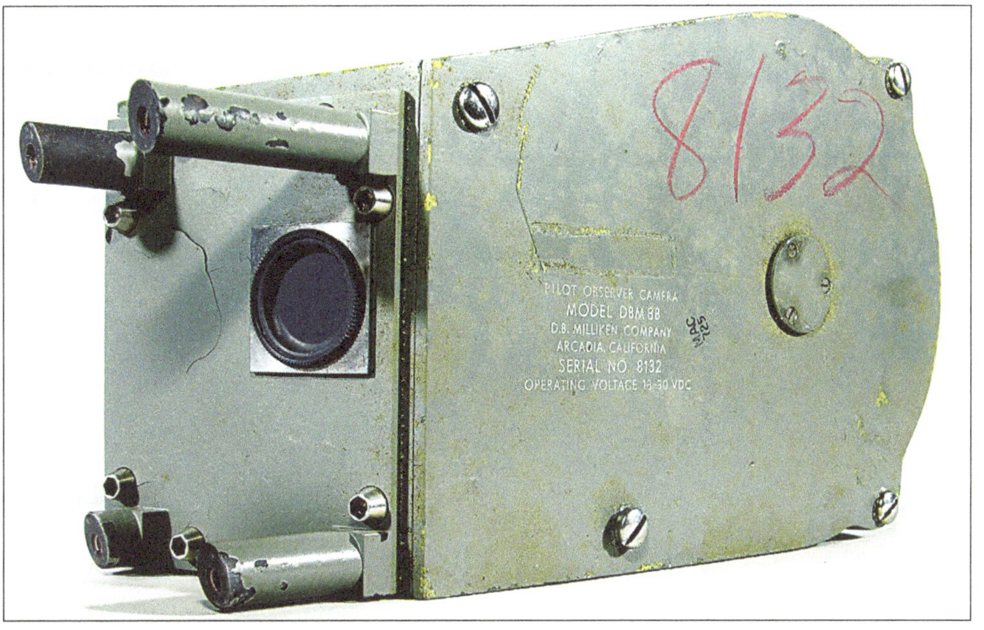

Figure 1.3. Mercury Pilot Observer Camera, NASM Cat. #A19790459000 (Smithsonian National Air and Space Museum).

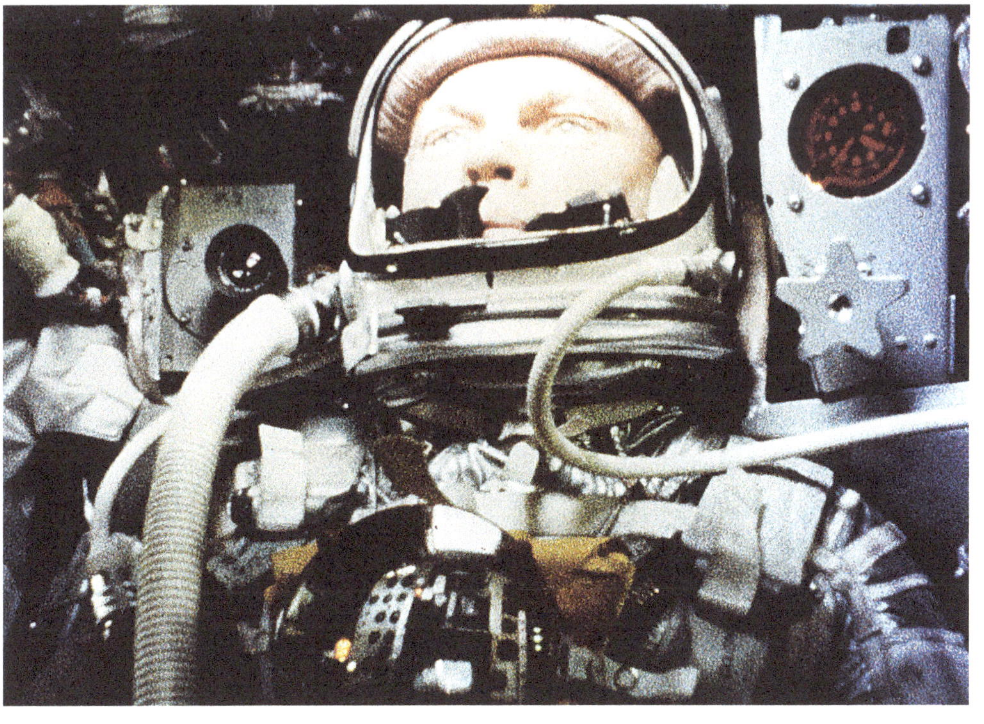

Figure 1.4. Instrument Observer Camera visible over Glenn's right shoulder and the chest mirror at the bottom shows some of the instrument panel, still frame from 16mm Pilot Observer Camera, *Friendship 7*, February 20, 1962 (NASA).

to listen to the astronauts, experiment, and slowly determine the best technology for science and engineering needs. This iterative process enabled personal photography through Project Mercury and beyond with engineers gathering input from a variety of sources to find technologies and methods in a program meant to inform later work.

While the missions of America's first two astronauts, Shepard and Grissom, were too short to accommodate handheld photography, the orbital flights of the next four astronauts began to provide NASA engineers with a sense of what astronauts required to return professional-quality photographs for a vast set of needs. For point-and-shoot daytime photography, John Glenn used an Ansco Autoset camera (the American brand name for the Minolta Hi-Matic 35mm camera) and inspired the trend of astronaut handheld photography. Glenn, after discussions with technicians and photographers, took it upon himself to procure a camera he thought capable of being manageable in space. As the story goes, following a routine trip to his favored Cocoa Beach barber, he wandered into the nearby drug store that sold small cameras. The Marine pilot settled on a $45 35mm camera, perhaps the most common camera type available commercially at the time, which he figured would simplify operating it in the small capsule (Figure 1.5). His three-orbit *Friendship 7* mission in February 1962 also included a NASA-selected 35mm Leica 1G camera (Figure 1.6), which engineers modified with a spectrographic lens and reticle for ultraviolet photography of Orion—making this the first human-operated astronomical experiment in space.

To make operating the Ansco easier, RCA contractor Roland "Red" Williams added a pistol grip handle and trigger to the camera so Glenn's spacesuit gloves did not hinder his work, permitting one-handed operation. Placement of the handle to align with the Ansco's exposure button required Williams to flip the camera upside down and move a flip-up Polaroid eyepiece to what became the top of the camera so Glenn could properly sight the Earth's horizon for the photographic study. NASA engineers also added a large reticle with bumper to the Leica, as the model selected had no built-in viewfinder and Glenn's spacesuit helmet prevented accurate aiming necessary for such precise photography. He could simply rest the reticle against his closed visor to steady it while pointing at Orion.

In Glenn's discussion of his flight photography in a June 2011 interview, camera choice and modification were apparently at his discretion since most at NASA found photography to be a diversion from the important engineering work Glenn was to perform on *Friendship 7*:

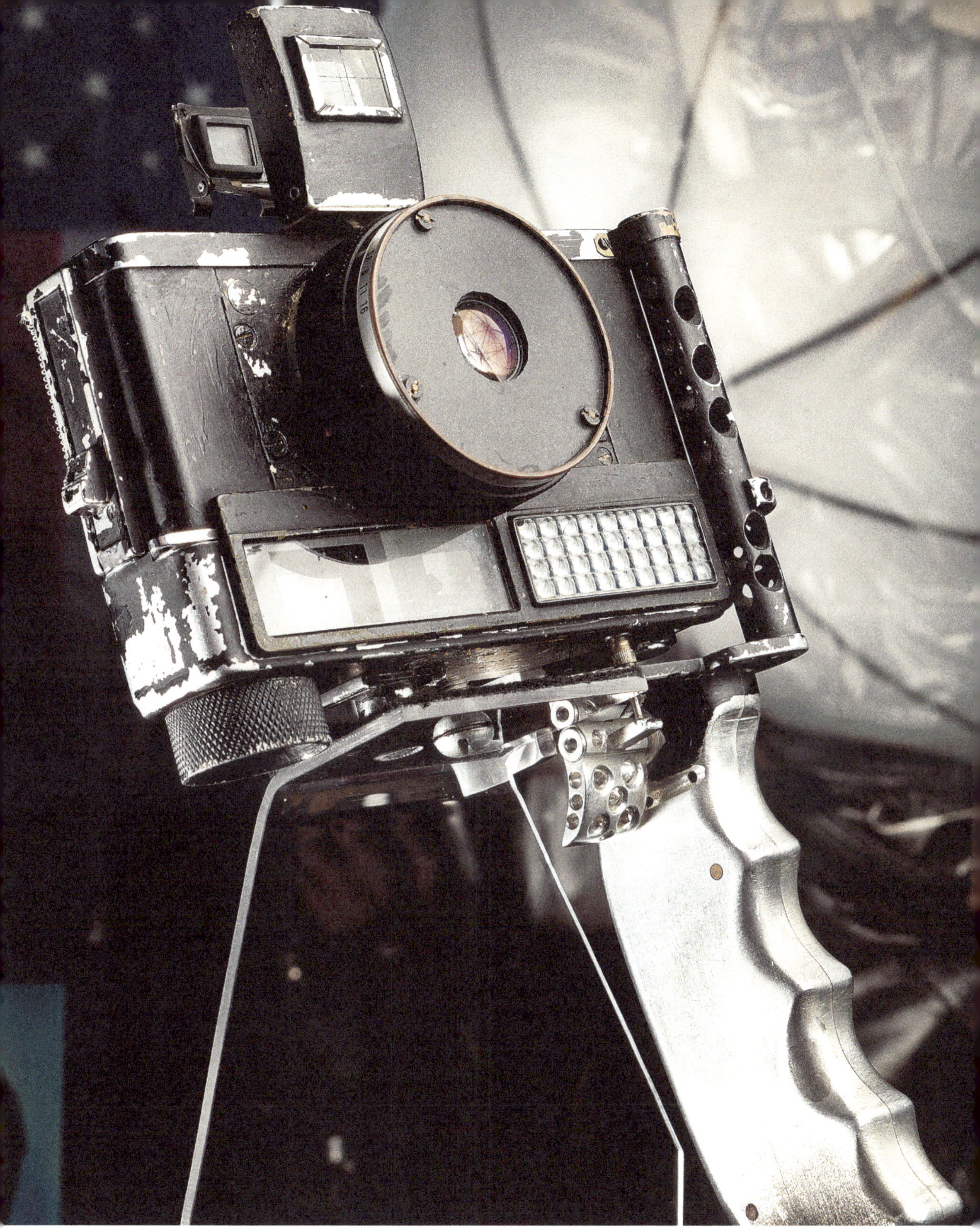

Figure 1.5. Ansco Autoset camera, *Friendship 7*, NASM Cat. #A19670198000 (Smithsonian National Air and Space Museum).

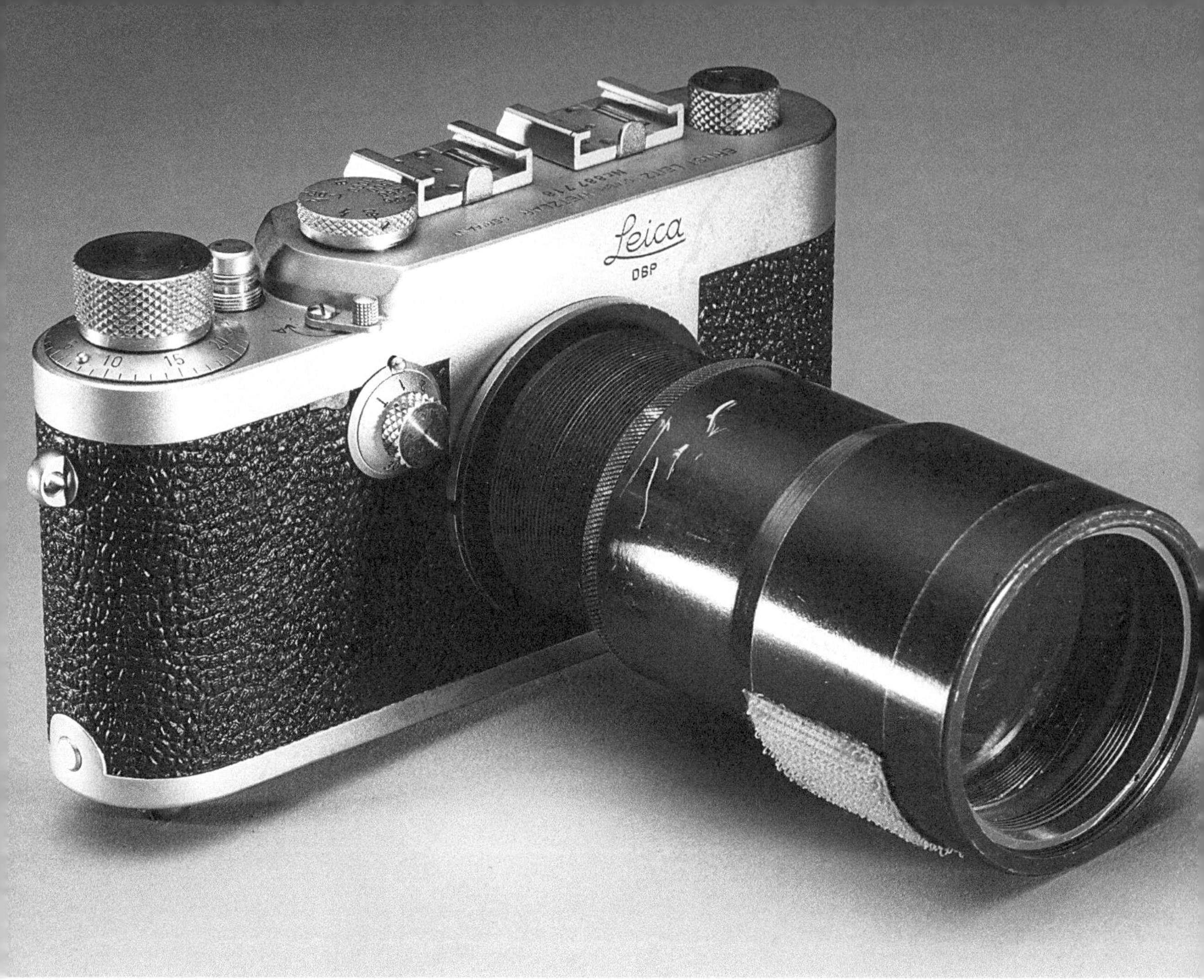

Figure 1.6. Leica camera, *Friendship 7*, NASM Cat. #A19670197000 (Smithsonian National Air and Space Museum).

Glenn: And the decision had been made that they didn't want to distract the astronaut.

Levasseur: So it was about distraction then?

Glenn: Yeah, that was the first thing. And so there wasn't going to be a camera onboard. And I talked to Bob Gilruth about it, talked to him a couple of times. He thought they'd gone a little far about that one also. So, he finally decided . . . to have one.[13]

The selection of two cameras indicates an awareness of potential technological problems that could arise when selecting equipment for spaceflight: modifications were possible to a point, but there was not one camera with the capability of serving all audiences for space photography. Glenn explained in our interview that "the one I was to use for regular shooting and just targets of opportunity, this [the Ansco] was it." With significant weight restrictions and close confines inside the capsule, carrying one camera for standard Earth orbital images and a second for astronomical observations indicates the competing interests even at this early stage of human spaceflight. Together, with all of their modifications, the two cameras were evidence of the need to find an all-purpose camera with interchangeable lenses and magazines to streamline training and mission time spent preparing the equipment.

NASA's camera technology experimentation continued with Carpenter's three-orbit mission in May 1962. For *Aurora 7*, they selected a Robot Recorder 36 camera, a model known mainly for its industrial and documentary application. Flight transcripts and the testimony of Christopher Kraft, NASA flight director at the time, indicated that Carpenter frequently operated the camera during his three orbits, and Kraft attributed the 250-mile off-target landing in part to the distraction of taking photographs.[14] John Boynton and E. M. Fields, authors of the official mission report section on spacecraft and launch vehicle performance, explained the concern a bit more tactfully than Kraft's chastising of Carpenter: "Astronaut Carpenter exposed an extensive series of general interest color photographs of subjects ranging from terrestrial features and cloud formations to the launch vehicle tankage and the tethered balloon."[15] The modifications to the camera to make a long series of exposures are obvious: engineers permanently attached a large film magazine to the camera, making a roll of around 200 frames available during the mission. Carpenter, so dedicated to his photographic work, valiantly saved the Robot camera from being waterlogged during his spacecraft egress—though his 16mm observer camera films were not as lucky and were nearly useless in post-mission analysis (Figure 1.7). The Robot camera, however, made no particular impact at NASA among the astronauts or engineering team, and this particular model never saw use again in space.

Scott Carpenter's part in the story of astronaut photography appears minimal at first, particularly since he only flew in space once. The somewhat notorious nature of his off-course landing, however, and the probability of photography's role in the miscalculation influenced later astronauts and their attention to flight plans. No other landing in the space program was off by such a great distance, and no other

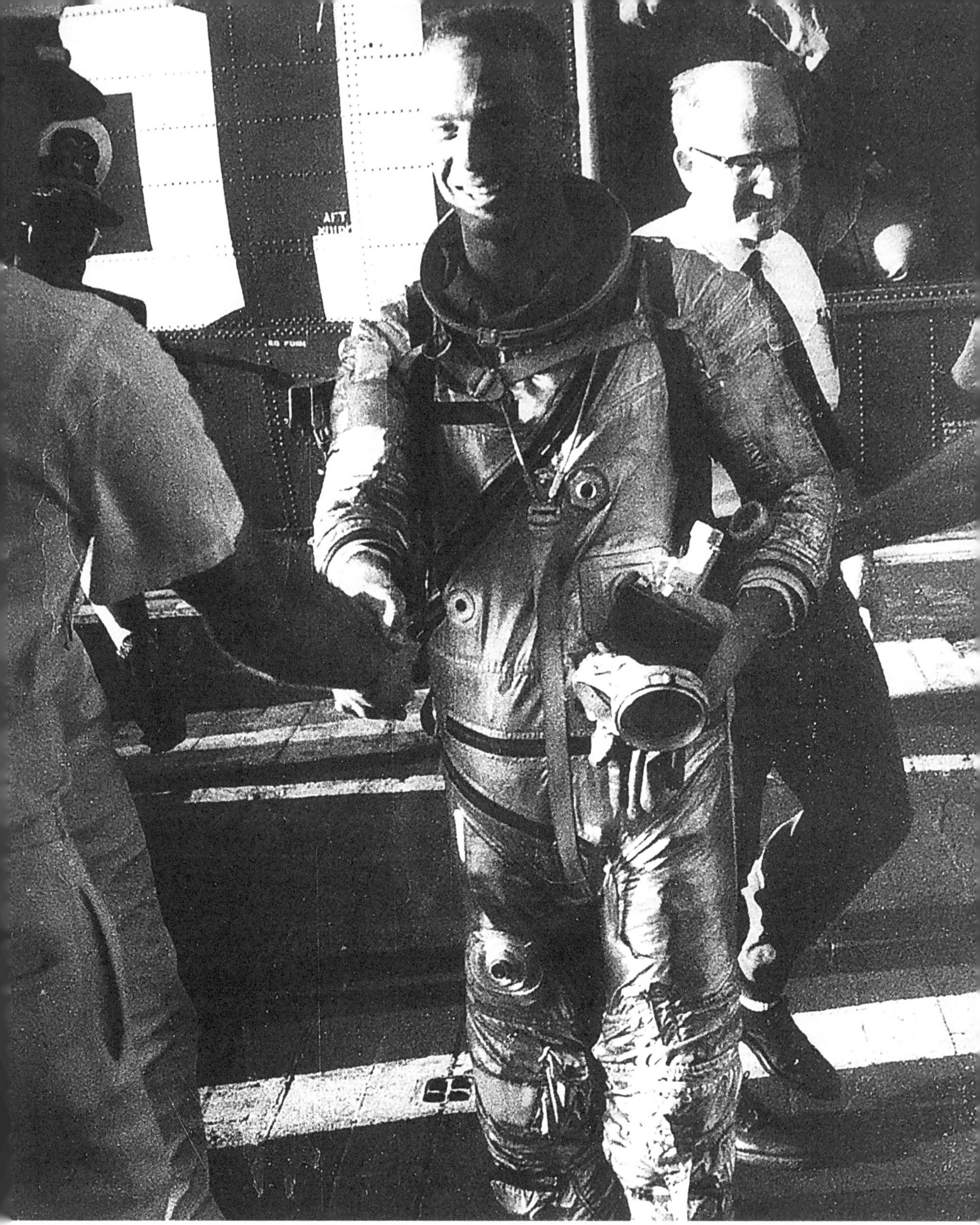

Figure 1.7. Astronaut Scott Carpenter on the recovery ship U.S.S. *Intrepid* after the *Aurora* 7 flight, carrying his Robot camera and spacesuit gloves, July 10, 1962, S62-04027 (NASA).

astronaut carried his handheld camera off the spacecraft and onto the recovery ship. Carpenter, still quite proud of his photographic work, fondly recalled the Robot Recorder in our interview: "And I remember that it was spring wound and that it had a five frames per second capability and 250 frames . . . big, big roll of film." He had opportunities to depict what Glenn had little time to capture on his flight. Glenn managed to bring back about seventy exposed frames on separate rolls of film, compared to the over two hundred photos Carpenter took on one roll. Working with Kodak on this challenge later, NASA sought a thin-base emulsion that could maximize the frames per roll and minimize the number of times astronauts needed to change film rolls. The massive film roll relieved Carpenter of the hassle Glenn found with separate rolls floating in the spacecraft, and perhaps opened up the opportunity for more photography than was practical in this early period of spacecraft research and development.

Wally Schirra hoped to take a high-quality professional camera on his *Sigma 7* mission in October 1962. Schirra's familiarity with cameras and photography was extensive prior to his mission, having purchased personal cameras over the years. Schirra added to that collection when he purchased a Hasselblad 500C from a Houston camera shop prior to his flight.[16] The Swedish-made medium format camera model, introduced to the commercial market in the late 1950s, quickly gained the respect of photographers for its quality, reliability, and flexibility. With the shutter located in the Zeiss lenses and not in the camera body, photographers could easily swap out lenses for different type of shots. The magazine design also allowed for far simpler loading of film over the laborious process for a 35mm camera. It seemed, based on Schirra's respect for the 500C, the Hasselblad was an obvious choice for general astronaut use in space from his first suggestion of it in 1962.

At that early stage, however, astronaut photographic work came under the classification of "experiments," for the purposes of understanding the Earth's airglow layer, making weather observations, and beginning the process of ground surveillance. The official flight report for *Sigma 7* in October 1962 (the third orbital mission) states:

> A series of terrestrial color photographs were taken by Astronaut Schirra for two purposes: (1) to aid in building up a catalog of space photographs of various physiographic features of the earth, such as folded mountains, fault zones, and volcanic fields; and (2) to obtain photographs of cloud patterns for comparison with those of other satellite programs.[17]

The use of images for public relations or anything other than scientific experiments was never mentioned in the reports, but images became, nonetheless, the way in which most people in the United States and around the world connected to the space program.

Astronauts, the Hasselblad Company, and numerous books and articles repeat this "origin" story of the relationship between NASA and Hasselblad as the brainchild of Schirra. In his autobiography and official NASA oral history, he explains his consultation with famous photographers at major pictorial magazines such as *Life* (Ralph Morse and Carl Mydans) and *National Geographic* (Dean Conger and Luis Marden) to find out what the most appropriate camera would be to achieve sharp and unparalleled photographs. Taking their unanimous advice, Schirra "decided that a Hasselblad, with its larger film frame, was more suitable than a 35-millimeter camera. I had a Hasselblad adapted."[18] Figure 1.8 depicts Schirra and Slayton examining two Hasselblads with engineer Red Williams, who did the modifications on Schirra's camera, as well as many others.[19] The Hasselblad lying on the table may be Schirra's own unmodified camera (the silver edging is a classic feature of these cameras), and he is handling what the caption states is the Hasselblad modified for flight, which appears darker and without a focusing hood. Schirra continued to credit himself and experienced photographer friends with determining what modifications would make the camera flight ready, explaining in his NASA oral history that those changes were made by a Pan Am laboratory near Cape Canaveral.[20] Schirra, in these interviews at least, leaves little doubt of his perception of the role he played in moving Hasselblad from its well-regarded role in the professional photography world to being the high-profile supplier to the glamorous visual world of spaceflight.[21]

Not even space heroes work in a bubble, and a decision by committee or small group such as the one pictured in Figure 1.8 is the most likely explanation of Schirra's ease in overstating his own role in taking the first Hasselblad to space. It comes as little surprise, perhaps, that during this period of technological experimentation, particularly in the realm of small equipment for astronaut use in flight, there was a willingness to test suggestions from trained pilots. These seven astronauts were surrounded (often hounded) by photographers. They worked with NASA's own photographers who documented every step along the way to space, so there was time to at least learn about what made for a good picture and what cameras those professionals had on hand. And, despite the

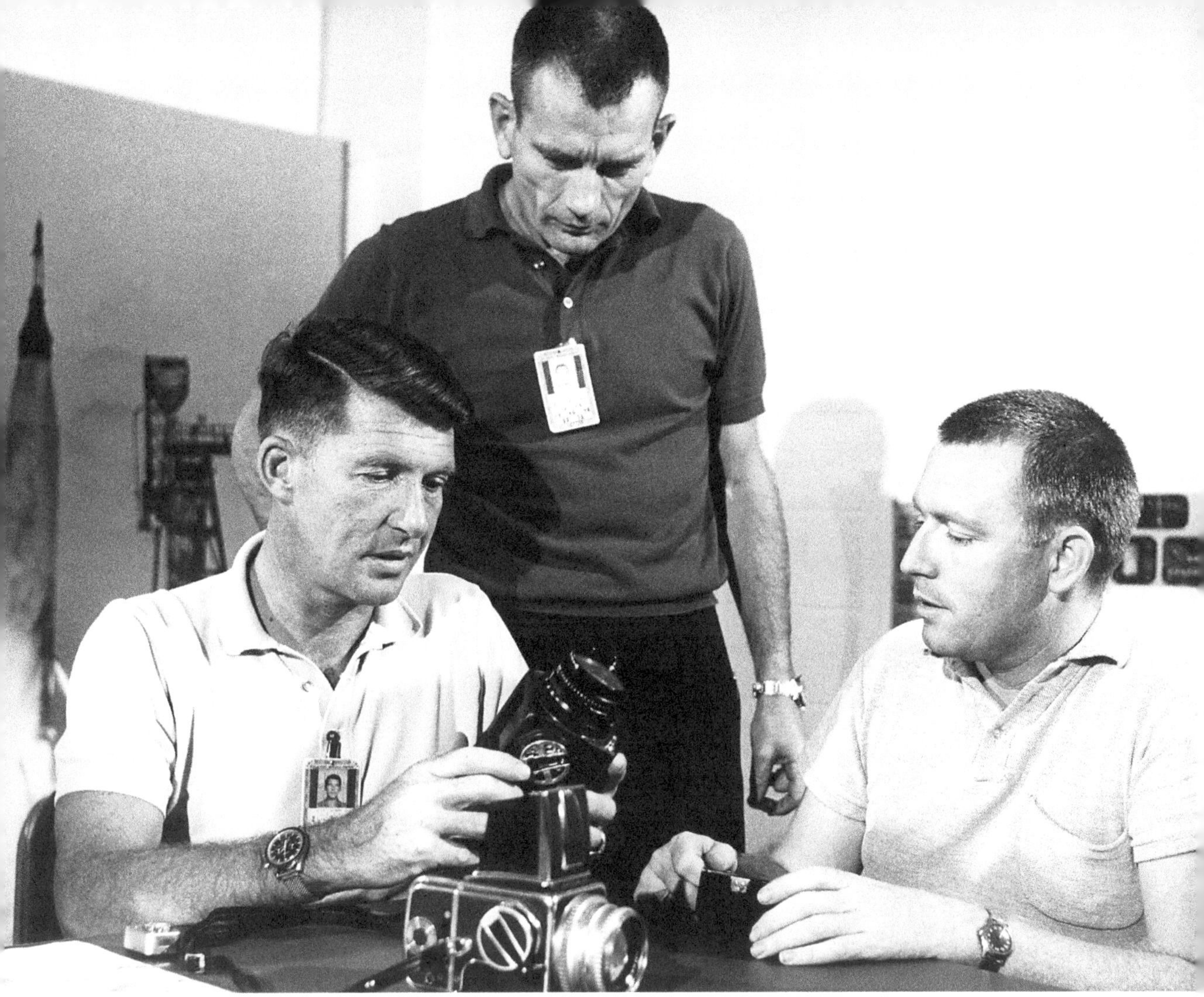

Figure 1.8. Astronauts Wally Schirra (left) and Deke Slayton (center) examine Hasselblad cameras with engineer Roland Williams, 1962, S62-06065 (NASA).

need for some necessary modifications due to the special environment of the spacecraft, engineers and astronauts alike came to agree with Schirra. From that point on, Hasselblad was the camera of choice on all successive U.S. space missions until the change to digital in the early 2000s (Figure 1.9). This relationship, perhaps the longest lasting between manufacturer and consumer in the space program, came to play a significant role in shaping the cultural legacy of both institutions.

In spite of a few experimental selections for handheld cameras during the first two orbital flights, Hasselblads were on board for the final two Mercury flights in October 1962 and May 1963. Schirra and Gordon Cooper were of one

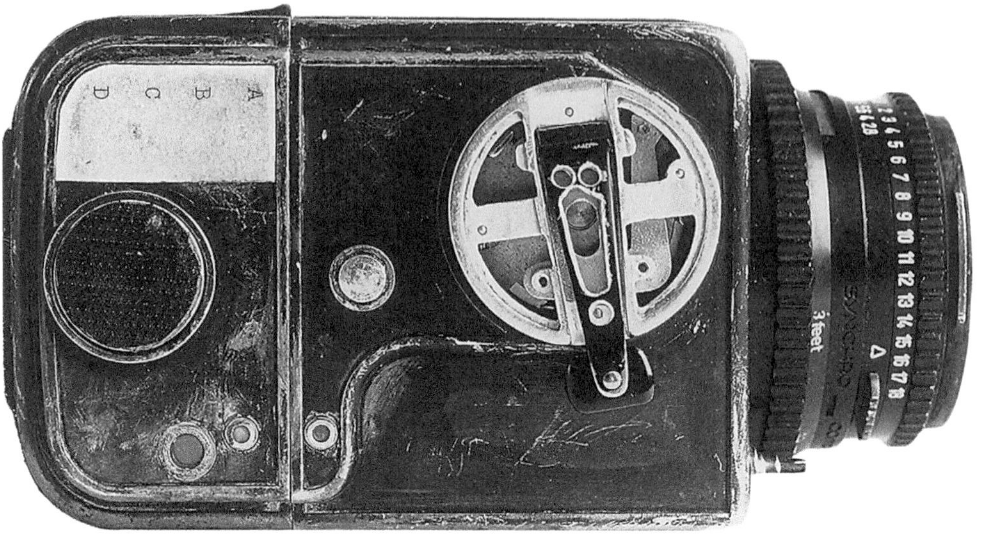

Figure 1.9. Hasselblad camera used by Wally Schirra on *Sigma 7*, October 3, 1962 (Courtesy of RR Auction).

mind in taking a professional camera into orbit—not an "instamatic" as Cooper characterized those used by Glenn and Carpenter.[22] Echoing Schirra's oral history and biography, Cooper tells of his lifelong interest in photography and dedication to taking professional-quality photographs during his spaceflight in his autobiography, *Leap of Faith*. In his characteristically sly tone, he explains, "As I kept shooting pictures, I didn't say anything and stayed quiet in the cabin, hoping the guys at Mercury Control would think I was asleep."[23] Cooper's attitude here indicates he noticed the reaction to Carpenter's overuse of the camera and planned ahead of time to keep his own extensive photographic work concealed from his superiors.

Other than a modified Hasselblad 500C, Cooper also carried a Robot camera on his Mercury mission (Figure 1.10). Like Carpenter's Robot Recorder 36, Cooper's Robot camera (probably the Royal 36 model shown in a mission report pamphlet) was modified, but this time, it was nearly unidentifiable next to its commercial form.[24] Also worth noting here are the measures taken to strip most of these cameras of their brand identities. Other than the Leica carried by Glenn, the cameras have blank exterior surfaces. So as not to appear to favor any particular brand, NASA did its due diligence as a government agency by decommercializing these technologies. Another notable modification to Cooper's Robot camera, described in the official mission report, are the three red "feet" attached to the right side of the camera. When used in space, Cooper placed the feet against the window over his head to get a steady shot.[25] While the exact need to relocate the film winding mechanism to the bottom of the camera is unclear, the camera was operated otherwise as it would be on Earth. A view of the camera's actual top shows the exposure button, exposure counter, a bracket for a flash or viewfinder, and a filled-in hole for the original location of the winding mechanism. The addition of a larger dial likely made advancing the film easier for Cooper while wearing his spacesuit gloves.

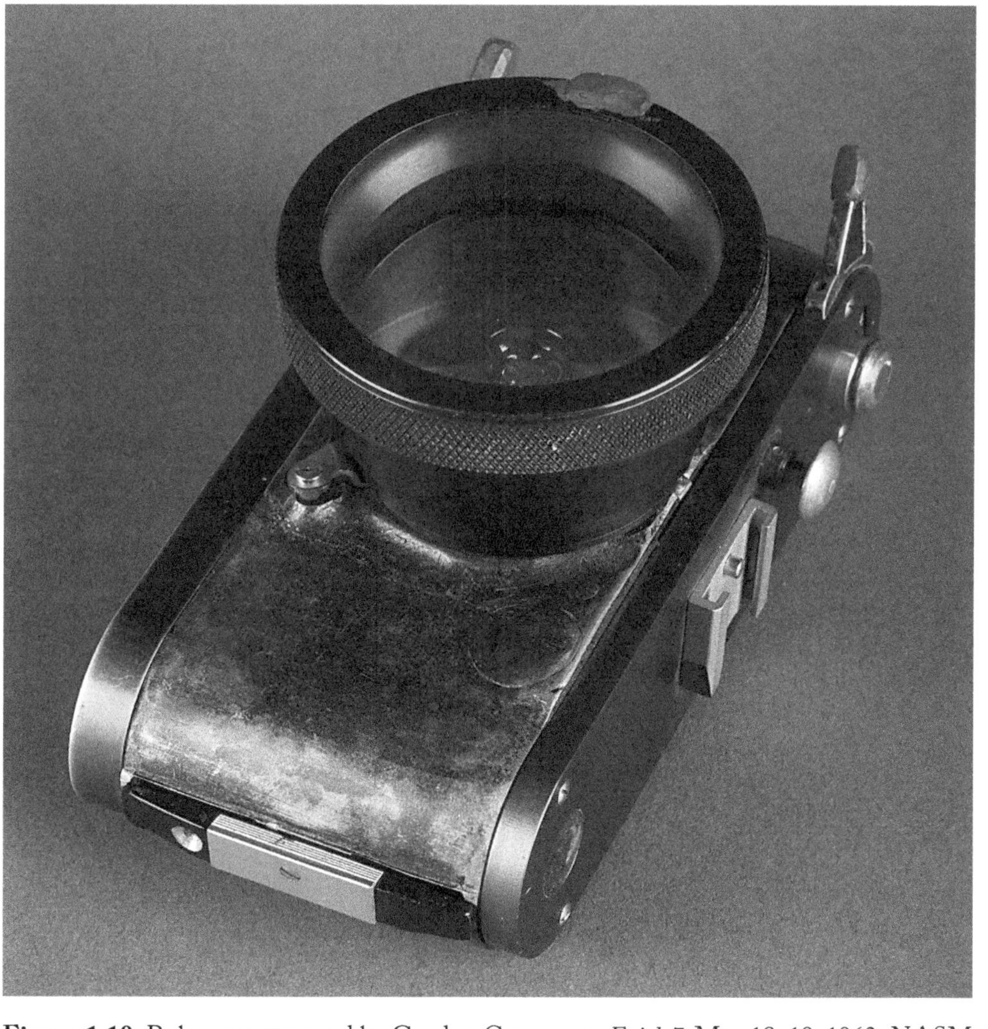

Figure 1.10. Robot camera used by Gordon Cooper on *Faith 7*, May 18–19, 1963, NASM Cat. #A19770553000 (Smithsonian National Air and Space Museum).

These early instances of selecting technologies for spaceflight seem more like chance and are evidenced only through anecdote, not through a government contract, purchase orders, or receipts. John Glenn in fact jokingly recalled that NASA never even reimbursed him for the Ansco camera.[26] Numerous popular articles, websites, oral histories, and books repeat the stories of the Glenn and Schirra camera purchases with no question as to the reality of the claims.[27] The heroic treatment of the original Mercury astronauts, and all astronauts really, leaves no doubt in the minds of most that the originators of photography at NASA were these high-profile personalities. This storyline seems all too easily manufactured. Like the tales of George Washington chopping down a cherry tree or other notorious stories of the nation's first president, historians are trained to question such simplistic stories of origination.

Without documentary or artifact proof, however, anecdotes of astronaut camera selection are hard to disprove. Even NASA's own photographic technology staff, particularly Richard Underwood, commonly repeated these stories, reinforcing rather than contradicting them by giving credit to existing NASA staff photographers, save those two instances. The astronauts themselves frequently credited two individuals from outside the astronaut corps with influencing their understanding of photography and cameras. Ralph Morse, *Life* magazine's photographer for the early days of the space program, not only photographed the astronauts, but also consulted with them about photography and cameras, earning the nickname of the "eighth astronaut" for his close relationship with them. The other was Bill Taub.

Chief photographer at NASA for decades, Taub participated in documenting early astronaut activities so closely that he was affectionately called "Two More Taub" for his constant requests to the astronauts for more photos.[28] Taub, an amateur photographer who began his career with NACA at seventeen years old in the early 1940s, was widely known for his artistic shots of the wind tunnels at Langley. He was particularly fond of using a Leica, which he most certainly promoted with the astronauts considering his admitted belief in its superior quality.[29] First employed as a model maker and draftsman, his artistic background, and apparent inability to mesh well with the Photographic Laboratory staff, put him in a position to work nearly independently on photographing the Mercury astronauts, shooting with little to no direction from his superiors.[30] Taub's lasting impact on the memories of Mercury astronauts conveys a sense of closeness and trust between the photographer and subjects. Taub, along with respected photographer friends from *National Geographic* (probably Dean Conger and Luis Marden, who Wally Schirra also cited as photographer friends), formed bonds with the astronauts as they all continued to photograph some of the more intimate and personal moments the first seven astronauts experienced in their preparations for spaceflight. While it can only be inferred from these references, professional photographers must have played some role in preparing astronauts Glenn, Carpenter, Schirra, and Cooper for their camera work in space.

While Mercury astronauts did communicate with notable photographers and NASA staff photographers during their training, it appears to have amounted to no more than camera operation tips and general thoughts on photography at the earliest stages of handheld cameras for still photography in space. Time allotted to photography in mission planning was minimal, so the public and scientific audiences for images happily accepted any images offered as evidence of the spaceflight experience. The lasting impact of experienced professional photographers on the space program comes in motivating the astronauts to take photography as more than just a science experiment. And despite the objections of NASA managers at the time, photography became a crucial task during missions, providing the rich visual record of human space exploration that people rely on for their memories of the early U.S. space program.

HASSELBLAD: PROFESSIONAL CAMERAS FOR PROFESSIONAL ASTRONAUTS

For a program that touted its "all-American" character, using an international source of technology seems puzzling, and a noteworthy exception to the standard procedures for U.S. government procurement. Common American-made camera brands available on the commercial market at the time included Ansco, Argus, Graphlex, and Kodak, among an even wider range of internationally produced brands, but something about Hasselblad made it the mainstay of the U.S. space program from Project Mercury though much of the Space Shuttle program. It is important to explore not only the history of the Hasselblad brand and its decades-long connection to NASA, but also the impact such a choice made on the images returned from space. How did the experiences of astronauts Schirra and Cooper make it so easy for NASA to overlook other (American) camera manufacturers in favor of the Swedish Hasselblad? What did each entity gain by making concessions on the production of the cameras that resulted in such a lasting association? This relationship represents an enduring and significant technological choice NASA made in the early days of the space program.

No serious scholarship is devoted to exploring the relationship between the aerospace industry and contractors for photographic technology, and relatively few scholars even mention it as a technological element of the space program.[31] In the case of NASA and its selection of Hasselblad, even less exists to substantiate the popular works that chronicle this Swedish-American partnership or the cultural significance of using non-American technologies to depict space exploration by humans.[32] On its own website, the Hasselblad Company spends considerable space on explaining its own history and the formation of its most fruitful partnership.[33]

The Gothenburg-based Hasselblad family began importing photographic equipment and supplies in the 1840s. Arvid Viktor Hasselblad, son of the founder of the family business, happened to meet George Eastman while Hasselblad was touring London on his honeymoon. The fortuitous meeting for the Hasselblad family evolved into a business arrangement that allowed the Hasselblad's photographic supply chain to become the sole distributor for Kodak products in Sweden for over eighty years. This relationship was advantageous for both sides: it allowed Kodak a simple access point to supply the growing numbers of Swedish photographers as technology got easier to use, and it allowed Hasselblad to expand its business to the U.S. market. Based on a simple handshake between Eastman and Hasselblad in 1888, the companies linked their business efforts, allowing Hasselblad easy access to the American camera market when their commercial manufacturing division began work after the close of World War II.

Arvid's grandson, Victor, traveled the world to learn about photography, in particular from his grandfather's American friend, George Eastman. Eastman mentored Victor in film processing and other specialties, and sent Victor on a worldwide expedition to meet with other companies and learn their time-tested manufacturing techniques. His knowledge of photographic techniques became known around Sweden and caught the attention of the Swedish air force during World War II. When it recovered a downed German reconnaissance aircraft, the air force enlisted Victor to help them understand the camera onboard.[34] In hopes of beginning their own photographic reconnaissance work, the air force paid Hasselblad to develop a camera for them, eventually known as the HK 7, the very first Hasselblad-built camera.

Victor's company soon grew to twenty employees, many of whom were auto mechanics and local watchmakers. Their knowledge of metals and precision work instantly gave the cameras a level of respectability for the intricate mechanisms, which also made them quite fragile. The Swedish air force, however, purchased 342 of Hasselblad's cameras from a product line that included the small HK 7 and the SKa4, a model specifically for mounting inside a reconnaissance airplane.

Just three short years after the end of the war in Europe, Hasselblad had an operational, commercially viable camera, and quite naturally to Victor, it was announced where he knew it would gain the most attention: at a New York City press conference. That old connection with the Eastman family and the Kodak Company left the door to the American market wide open, and the first commercial model, the 1600F, made a splash among professional photographers (Figure 1.11). It would not take long for Hasselblad and Kodak to come together again, but this time in space.

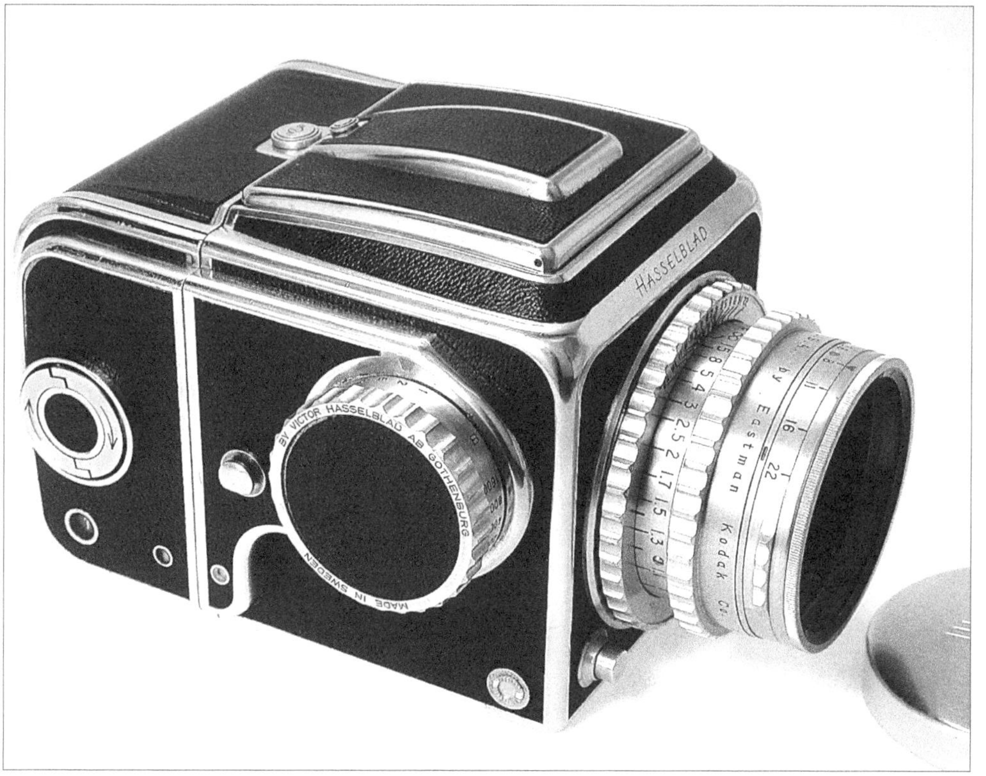

Figure 1.11. Hasselblad 1600f model (Richard Nordin).

Introduced in 1948, the 1600f was a single lens, mirror reflex, 6 x 6 cm medium-format camera of high quality and precision, but quite delicate. Hasselblad did ensure, however, flexibility of use by designing the camera with the options of changing lenses, viewfinders, and film magazines very easily. To overcome the 1600f's delicate nature, the 1000f entered the market in 1953, improving on many of the former's flaws. Leading up to its release to the public, *Modern Photography* undertook extensive field tests of the camera, proving its durability and reliability.[35] The magazine's staff ran nearly five hundred rolls of film through a single camera, and dropped it twice, without finding a single problem. With this success, Hasselblad began an intensive media campaign through popular professional photography magazines to extol the virtues of the 1000f. The Supreme Wide Angle (SWA), 500C, and Super Wide Camera (SWC) models soon followed.

This is where the stories of camera manufacturer and fledgling space program meet. Not long after the SWA (1954), 500C (1957), and SWC (1959) models entered the market, Schirra and NASA's own professional photographers took notice of the respected Swedish-made camera. The Hasselblad Company's operating philosophy seen in advertising campaigns showing their cameras having "modularity, versatility, and reliability" became quite attractive to anyone seeking equipment that needed to function flawlessly in a relatively untested and unknown environment. Their advertising campaigns highlighted this mantra and spun it in a way that made the appeal of the brand to NASA quite clear (Figure 1.12). Marketing materials featured various models, configurations, and the overall flexibility they offered customers, commenting on features in line with what NASA sought, or what the advertisement notes, "Few amateurs have the time to master this science so thoroughly." Later, Hasselblad played up their success and attentiveness to NASA's specialized needs to advertise themselves as the most flexible camera manufacturer (Figure 1.13). The second ad leaves the options open to the buyer, a role rarely given to personal technology consumer before the age of personal computers and software options.[36]

With the wide availability of quality 35mm cameras in the early 1960s, one might ask why NASA focused so quickly on the medium format 70mm cameras that became a staple of human spaceflight through most of the Space Shuttle program. In comments published as part of the SPIE Technical Symposium in the summer of 1964, John Brinkmann and manager John Eggleston offered one reason

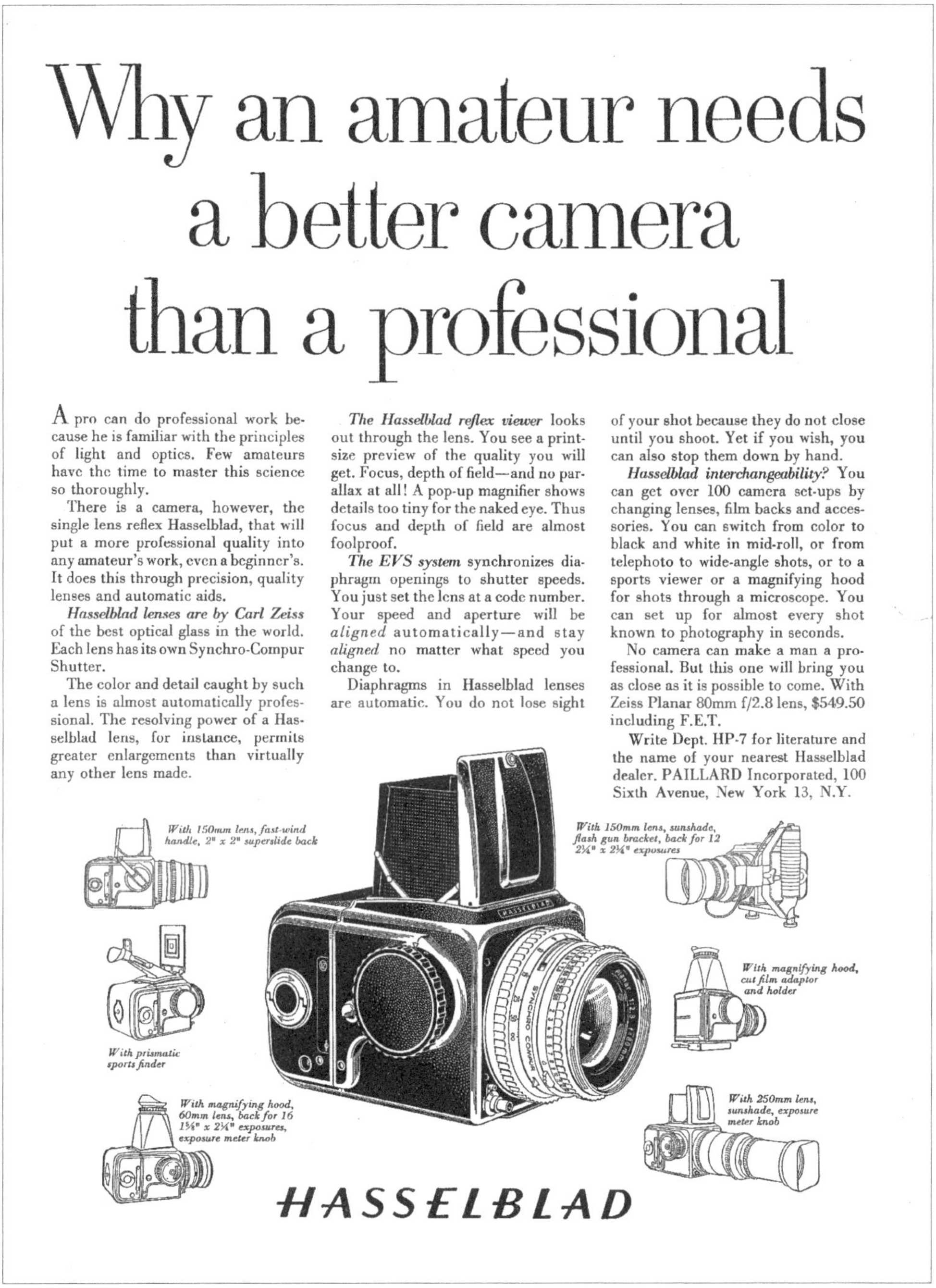

Figure 1.12. Hasselblad advertisement, *Popular Photography*, July 1961 (Courtesy of Smithsonian Libraries, Washington, DC; courtesy of Vicktor Hasselblad AB).

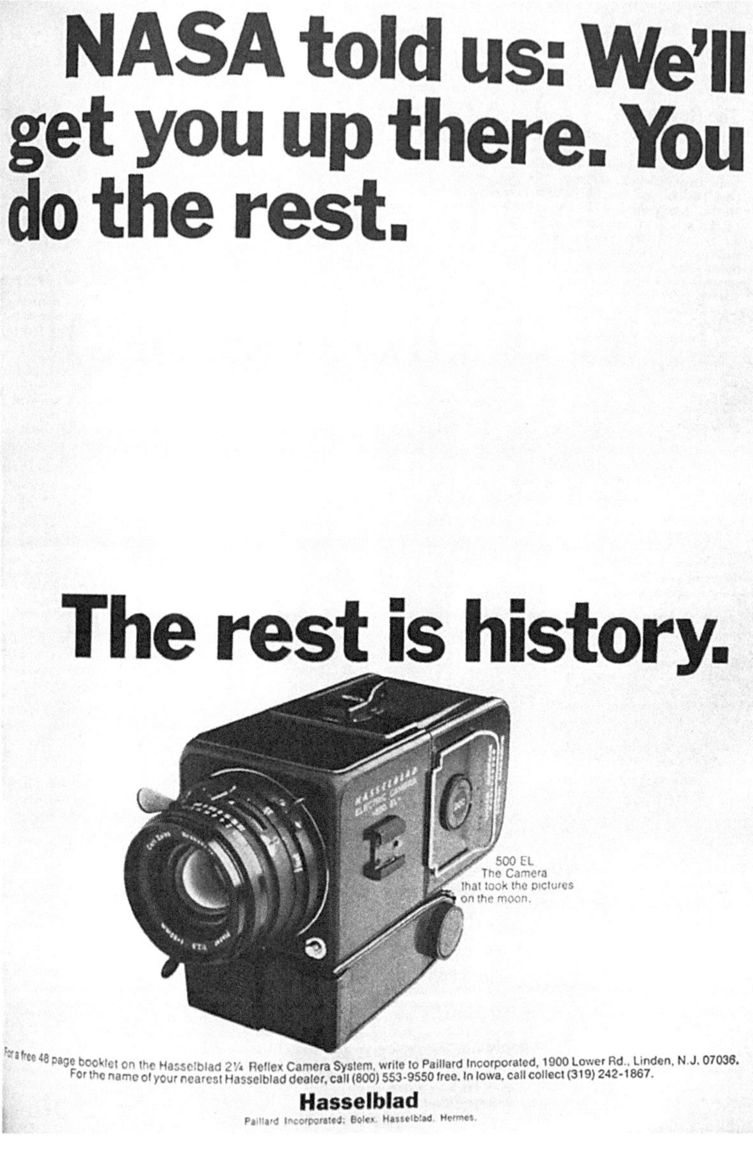

Figure 1.13. Hasselblad camera advertisement, *Newsweek*, August 11, 1969 (Courtesy of Viktor Hasselblad AB).

for this quite early in their comments in reference to Mercury experience: "Glenn noted that the color fidelity of his photographs did not exactly match his memory of the sunrises, sunsets, and space fire-flies that he saw and photographed."[37] As expectations for scientific, technical, and public relations imagry understandably rose after that first flight, NASA responded by seeking out a more versatile camera capable of higher quality color photography.

Both NASA and Hasselblad had reason for caution as development of the cameras proceeded. The microgravity, low pressure, and varied temperature environs of space and technical requirements for spacecraft launches meant that the cameras required modifications. These problems included dealing with off-gassing, heat, weight, lubricants, and mechanical operations. Realizing the value of developing a relationship through such a project, Victor Hasselblad and other company representatives apparently took all necessary measures to modify their cameras for safe operation inside and outside a space vehicle. As the professional relationship (and personal as it turns out—Victor and his wife became good friends with a number of astronauts) developed, so did the manner in which the modifications occurred.

From the start, one of the more fashionable features of cameras—at least in the late 1950s—raised concerns at NASA. For Schirra's mission, technicians stripped the leatherette coverings to avoid any off-gassing inside the small Mercury capsule. This ad hoc solution during Project Mercury, a period of developing hundreds of new technologies, became easier when NASA worked directly with Hasselblad for Gemini and Apollo.[38] Engineers at Hasselblad would later manufacture all NASA cameras without such coverings.

While removing the camera's coverings minimized adhesives and other materials off-gassing into the spacecraft environment, minimizing camera weight required more creativity. Engineers at NASA and Hasselblad determined that some normally critical elements of the camera's structure needed to be removed or altered in order to make them as light as possible. In spaceflight, every gram of weight is counted because a proportional amount of fuel is required to lift that weight into orbit and beyond. By agreeing to construct the camera case using lighter metals, Hasselblad opened up its design and manufacturing process to significant change. Engineers determined that removing parts such as the mirror, focusing screen, hood, and case linings meant less weight and they were not required for space operation. As a helmeted astronaut could not look through a small area to view the possible image, the mirror used to bounce the image from the lens to the viewfinder was unnecessary, as was the viewfinder itself. To make up for the loss of the viewfinder, part of the astronaut training program included how to aim the camera without such help.

Operation of a camera in space turned out to be a slightly different process than on Earth, and that unique environment forced engineers to find a new way to lubricate internal mechanisms. NASA found that Hasselblad's recommended lubricants for consumer cameras, Isoflex Topas L 32 (grease) and Isoflex PDP 48 (oil), would evaporate or solidify in the vast temperature swings of space. The use of an alternative, a synthetic material requested by NASA, meant that the astronauts did not need to maintain the cameras during the mission, and minimized the risk that cameras would jam during operation. It was not a perfect process, but vacuum chamber testing meant that the cameras were prepared as close to the right configuration for space operation as possible.

To minimize the influence of the sometimes-high temperature environment of space and of the lunar surface, Hasselblad switched the black metal camera body to silver to reflect more of the sun's heat. Temperature swings on the Moon are vast, going from 120° C during the day to minus 150° C at night. By using aluminum-covered or silver painted surfaces on equipment for the lunar surface, NASA reduced potential heating problems with cameras and scientific experiments.

For missions lasting from days to over a week, and with the intent to maximize the available frames of film while minimizing weight, NASA needed thin-base films with need-specific emulsions and large-capacity magazines astronauts could swap out at a moment's notice. To save weight during use of the Hasselblads on late Mercury and early Gemini flights, NASA contracted with Cine Mechanics in California to lighten the film magazine construction to save weight and increase internal volume for film. The same company later added additional metal to the magazines to provide an increased measure of radiation protection when the spacecraft left the Earth's magnetic field to go to the Moon.[39] The space agency, to its benefit, could also immediately tap into their existing relationship with the country's largest photography product manufacturer, Kodak, to create anything deemed necessary for spaceflight photography. Kodak's long-standing relationship with Hasselblad made the process even simpler, allowing the three to use their collective experience to problem solve for film needs going into the Apollo lunar missions in particular. As early as 1964, NASA contracted with Kodak and others to study the potential effects of light reflection off the lunar surface on different film types proposed for use by astronauts, so the relationship went beyond just ordering special films.[40]

Finally, the dials, switches, and buttons on the camera underwent modification. Enlarging them made operation by astronauts easier when wearing bulky spacesuit gloves. This rather extensive set of modifications to such well-respected equipment proved Hasselblad's responsiveness to what they acknowledged was an outstanding means for promoting their products, but also of establishing their brand as part of an historic American program.

Long touted as an American program based on American ingenuity and technology, NASA's human spaceflight efforts notably passed over American camera manufacturers as well as those from Japan and other parts of Europe in favor of this relatively recent addition to the camera market. For decades, companies manufacturing lenses in Germany and France were known to employ the best optical engineering methods to create products respected around the world. In selecting Hasselblad, NASA accumulated technical information and anecdotal evidence from the Mercury flights that informed their decision to focus on the Swedish brand above others. With Schirra's endorsement and the interchangeable components of the 500C, NASA had little reason to look at other options when it needed to achieve so many objectives with as few pieces of equipment as possible.[41]

Restrictions did (and still do) exist, however, for the purchase of products by the U.S. government from international companies.[42] Conveniently, NASA worked around this issue by proving through testing that the Hasselblad was a superior product for the tasks required (an exception allowable through the law), and by working with a purchasing agent in the United States, Paillard, Inc. of Linden, New Jersey, so that it became an indirect purchase that still supported American businesses.[43] But for scientific, engineering, and promotional purposes, NASA could not afford to use unproven or unreliable technology to record each unique mission. They simply selected the best available technology to produce some of the most iconic images of the twentieth century.

PHOTOGRAPHY COMES OF AGE ON GEMINI

Gemini program decision-making about cameras was more rigorous than it was in Mercury. Longer missions, an additional crewmember, and more storage space allowed crew equipment engineers at NASA to research both off-the-shelf and specially developed cameras for the increased photographic work prompted by those stunning Gemini IV results. Managing the research and responsibilities among a larger set of interested offices at NASA meant increased organization and management so as not to duplicate efforts or expend funding needlessly. Yet by the final four Gemini missions in 1966, departments across JSC were researching the next generation of cameras astronauts would employ to capture the stunning images of Apollo. As with other equipment and methods used during Gemini missions, the selection of cameras was intended to solidify the understanding of the astronaut-equipment interface in space.

Camera equipment documentation reflects the efforts of managers and engineers to develop versatile equipment for use on multiple experiments. When putting the possibilities on the table, as scientists and engineers did when selecting cameras for Gemini missions, they recognized the need to balance between flexibility and the constraints of the spacecraft, finding that, "use of an all-purpose camera to satisfy several needs on one flight necessitate building a larger, heavier system than any one experiment by itself would require."[44] And making anything for space larger or heavier was normally not an option. Getting experiments to conform to fewer pieces of equipment, and modifying those to satisfy more realistic experiment goals, may have resulted in some experiments being cancelled or changed, but it largely allowed interested audiences to gather the needed photographic data.

These discussions prior to crewed missions were held at levels far below the managers and reflect the thoughtful and collaborative push for extensive photographic work. NASA scientific and technical audiences voiced their needs and asked NASA management to recognize its value and get behind committing time and resources to photography. Photography from the first two missions alone provided the evidence engineers, scientists, and technicians needed to prove the rhetorical and scientific value of astronaut photographic work, justifying the additional time, money, and training needed.

Planning for increased photographic experiments developed long before missions took place, but documents are lacking in establishing the initial decisions about which cameras to employ for Gemini. According to a November 1964 memorandum, some five months before the launch of the first crewed flight, Gemini 3 (March 23, 1965), October and November meetings between interested scientists, NASA engineers, and Department of Defense (DoD) staff discussed common needs in order to minimize the photographic equipment required for each mission. Other than aligning experiments with available cameras, meaning the 70mm Hasselblad (Figure 1.14) and what NASA called the "MSC 70mm camera" (Figure 1.15), they found instances when experiment results would presumably align with PAO requirements: "It was concluded that the requirements for synoptic terrain and weather photography fitted well with the Public Information Office requirements for still photography."[45] Ultimately, cameras were a technology ripe for weight saving since simply changing lenses or film types could meet different needs.

While not ready for use until Gemini IX-A in June 1966, the "MSC 70mm camera" made by the J. A. Maurer company of Long Island, New York, became one of only two non-Hasselblad still cameras used during Gemini (the other being a Zeiss Contarex model used on missions IV, V, and VII) (Figure 1.16). NASA planned on the Maurer 70mm for a variety of photographic experiments, for which it was better suited because of the lens types available from the company's subcontractors that permitted capture of low-light, ultraviolet, and infrared images.[46] Standard Hasselblad cameras had no such capability, and because a film format of 6 x 6 centimeters limited the angle of view, NASA opted to use the super-wide version of the Hasselblad to make interior photography and exterior images a bit more error-proof.[47] Unlike the technological experimentation during Project Mercury, NASA's selection of cameras for the enlarged capsule and extended missions of Gemini required additional consideration of efficient packing and timeline planning. According to Helmut Kuehnel in the Apollo Experience Report on photographic equipment used in this period, the Maurer 70mm camera never performed reliably in its scientific work.[48] Though expectations appeared to be high for a purpose-built 70mm scientific camera prior to Gemini missions, Hasselblads alone continued into Apollo.

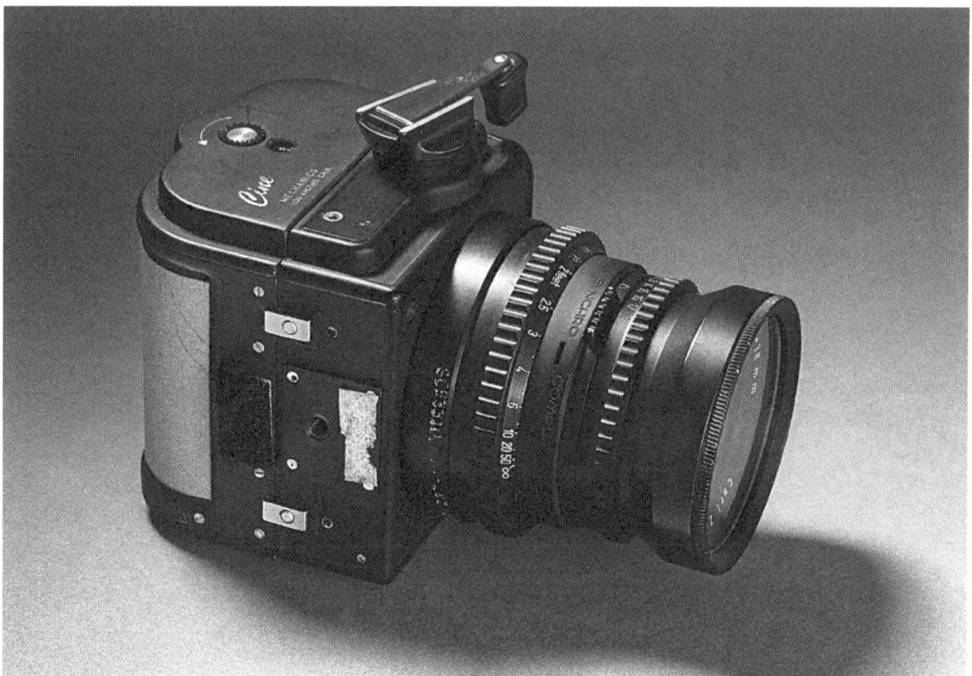

Figure 1.14. 70mm Hasselblad SWC from Project Gemini, NASM Cat. #A19790525000 (Smithsonian National Air and Space Museum).

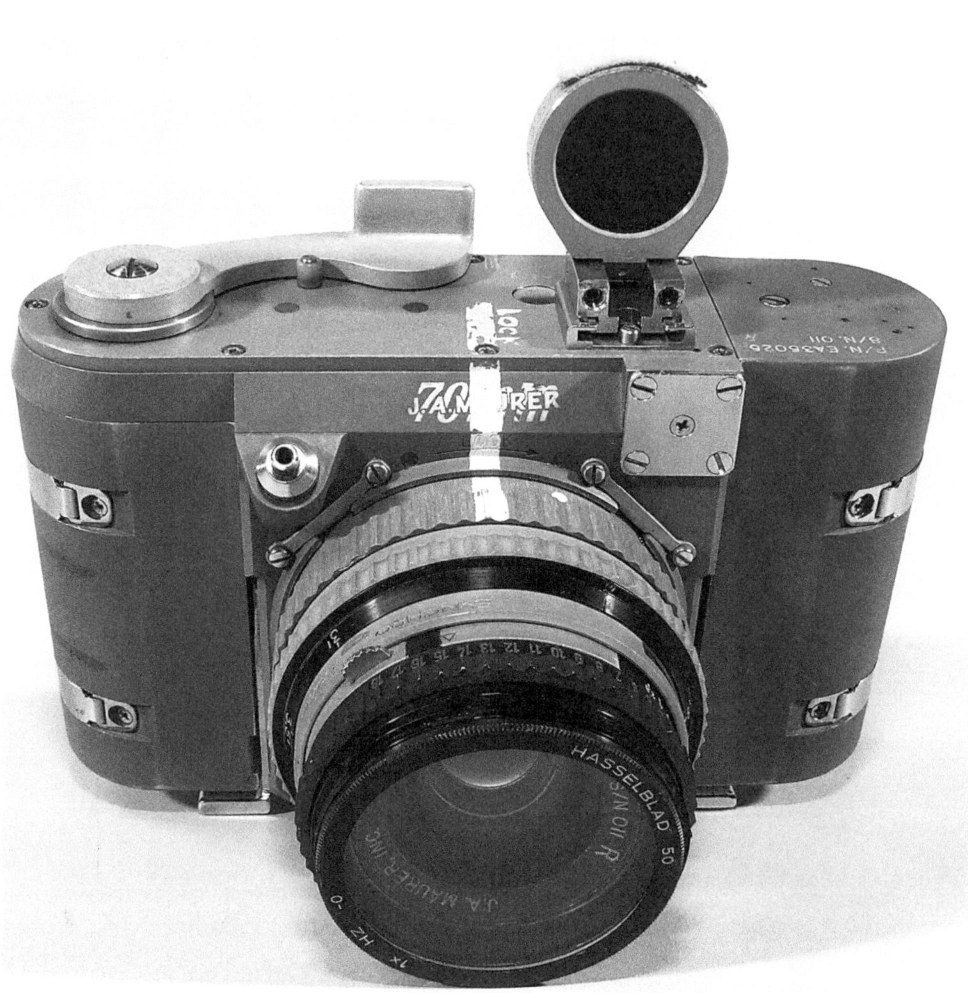

Figure 1.15. 70mm J. A. Maurer still camera from Project Gemini, NASM Cat. #A19761794000 (Smithsonian National Air and Space Museum).

Following Gilruth's reaction to the Gemini IV images, photographic efforts during missions did increase, and quickly. Press kit information for the next mission, Gemini V, with Gordon Cooper commanding and Charles "Pete" Conrad as pilot, shows the number of photographic experiments for scientific, technological, and military purposes doubled.[49] And while the press kits for Gemini IV and V reflect a repetition of experiments using nearly identical language, the mission time allotted to photographic work the astronauts undertook increased significantly, as did the secretive nature of their work. Some of the added photographic work for Gemini V came at the request of the military. According to mission commander

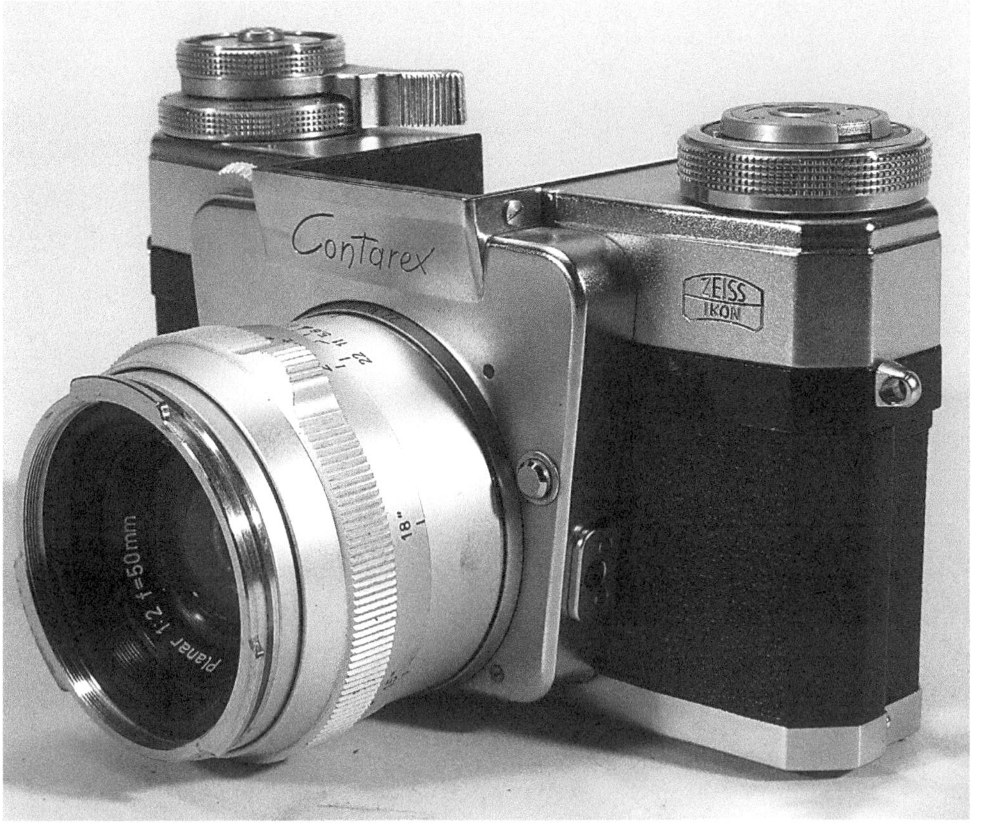

Figure 1.16. 35mm Zeiss Contarex camera, NASM Cat. #V19890085039 (Smithsonian National Air and Space Museum).

Gordon Cooper, veteran of the final Mercury mission, the increased workload resulted in more types of cameras in the spacecraft. In his autobiography, *Leap of Faith*, and his official NASA oral history, Cooper recounted feeling as though their small cabin was filled to the brim with nearly twenty different cameras, handing the astronauts responsibility for a tremendous amount and variety of photographic experiments.[50] In reality, there were only two cameras officially assigned to the mission, but far more lenses and film than on the first two Gemini missions or Cooper's Mercury flight. His perception of being overwhelmed with camera equipment illuminates the shift brought about by Gilruth's revelation regarding photography's value. What the press kits and mission documents do not specifically support, however, is how NASA planned to use the photographs collected. The assignment of photography experiment code numbers according to the type of project indicates that particular audiences existed for the images, but not what intermediary steps NASA took to review, select, and disseminate photographs for publicity purposes.

Training astronauts for mission work unrelated to spacecraft operation consumed far more time than one might expect from the lack of direct documentation on training. Gemini mission reports reflect a number of broad categories summarizing the preparations and experiences for each mission. Included in this post-flight documentation are descriptions of the briefings and training given for mission experiments, including the biomedical, optical, photographic, and scientific experiments. The breadth of these sections shows the degree to which crew time, as a portion of overall training, was spent preparing for experiments (Table 1.1).

We see quite clearly that experiment training decreased over the course of the Gemini program, while the number of experiments remained relatively steady but somewhat dependent on the expected mission duration. The proportion of photographic experiments to experiments overall remained relatively constant, going from zero on Gemini 3's short test flight to 66 percent of the work on Gemini VI-A. Nothing, however, indicates that photographic training for any of the experiments was proportional to the time spent in training since there are no records to reveal enough detail about training time to judge the relative priorities given to different experiments.

Table 1.1. Data compiled for Gemini mission training. Source: Project Gemini mission reports.

Mission	Mission Duration (days/hrs/min)	Avg. Training Time (hrs)	Avg. Experiment Training (hrs)	Experiments/ Total Training	Total Number of Experiments	Photographic Experiments
3	4h, 52m	411	200[51]	49%	3	0
IV[52]	4d, 1h, 56m	n/a	n/a	n/a	11	3
V	7d, 22h, 55m	457	150	33%	17	7
VII	13d, 18h, 35m	431	100	23%	18	3
VI-A	1d, 1h, 51m	332	23	7%	3	2
VIII[53]	10h, 41m	523	n/a	n/a	10	3
IX-A	3d, 0h, 21m	538	n/a[54]	n/a	7	2
X	2d, 22h, 46m	518	65.5	13%	15	5
XI	2d, 23h, 17m	541	36	7%	11	6
XII	3d, 22h, 34m	519	28.5	5%	14	7

For Project Gemini at least, management saw photography as an integral part of the time reserved for experiments. These projects, steered by technical staff at MSC, scientists, and the DoD offices, were designated based on their origin: "MSC" for those originating at the NASA center, "S" followed by a number if from the scientific community, and "D" followed by a number if requested by the DoD. In a report summarizing the defense experiments, only three of the sixteen requested and flown experiments were photographic in nature, all of which were only attempted on Gemini V.[55] Although the DoD continued with other experiment types on later missions, perhaps this lack of interest in continued photographic work was linked to the fact that photoreconnaissance satellites remained the preferred method for observations. The insertion of people between image collection and processing certainly complicated matters.[56]

As the first foray into integrating photographic plans and training into the daily routines of astronauts, Project Gemini focused considerable attention on preparation. Training consumed copious hours. Handling cameras and small equipment became more natural as astronauts trained and made repeat trips to space, but they also needed to divert attention to training for extravehicular work, rendezvous and docking, and intricate spacecraft maneuvers. What did not decrease was the volume of photography completed during the program. That, in fact, increased in its proportion of total experiments with the number of frames captured coming close to 500 during the Gemini XII mission in late 1966.[57] By that time, photographic information no longer came from experiments alone. Astronauts had more freedom to capture images referred to as targets of opportunity: general interest views or those containing particular information or experiences they wished to capture on film. This freedom to move beyond the bounds of experimentation and into a place where astronauts took some small measure of time to consider their photographic work shows a transition in their role from simply recorders to astronaut-photographers.

TO THE MOON ON APOLLO

With a goal as monumental as reaching the Moon in less than a decade, preparations for every aspect of the lunar program required attention even before Gemini got underway. Mercury astronauts and supportive engineers made handheld photography a part of astronaut performance and mission reporting, so it already existed as a formalized part of Gemini mission planning. Preparing for the unique character of Moon voyages, however, required special consideration for nearly every item astronauts needed. In early 1964, engineers from MSC's Advanced Spacecraft Technology Division (ASTD) and the Instrumentation and Electronic Systems Division (IESD) were already hard at work with photographers from John Brinkmann's Photographic Technology Division (PTD, later PTL for Laboratory) to discuss development of a lunar handheld camera. The seventeen-page statement of work and design document not only details precise requirements for a camera for use on the lunar surface, but also takes the somewhat surprising position of recommending a single stereographic camera for the entirety of astronaut lunar photography.[58]

The language used by NASA management in discussing the goals of photographic work focused on pragmatic issues of cost, reliability, suitability, modifications needed, and weight and volume considerations as seen in a May 1966 memo from Deke Slayton to Robert Gilruth.[59] Such discussions usually originated at lower levels of the organization, as did this one. And though they addressed seemingly mundane details, such as what manufacturers to choose and the extent to which cameras needed modification for flight use, the details were considered important at the highest level of management at the primary center for human spaceflight. Suggested in Slayton's memo to Gilruth was the formation of a committee to make decisions about camera selection and priorities, the Camera Development Review Board, which George Low, MSC deputy director, formally created on July 15, 1966.[60] The board's initial membership included Gilruth's special assistant, Paul Purser, as chairperson, plus representatives from the PTL (John Brinkmann), IESD (Myron Curtner), the Experiments Program (Fred Pearce), Flight Crew Support Division (FCSD, Helmut Kuehnel), Space Sciences (George Bonner), and Public Affairs (PAO, Andrew Sea). While Slayton's initial suggestion of the board did not include a public affairs staff member, Low's revision to include such a representative indicates his and upper management's awareness of the need for someone to speak for the agency's public needs.

Still, the primary responsibility for camera selection ended up in the hands of the IESD, despite Slayton's explicit objection that "operational requirements, human factors, crew training, and management, development, and procurement of crew operational photographic equipment" remain a responsibility of the FCSD.[61]

As the only dissenter to the late 1967 revision of the Camera Development Review Board's charter, Slayton made a claim on these tasks for his division. This likely indicates an ongoing difficulty or rivalry in communications between FCSD and IESD, a conflict clear from the initial meeting of an ad hoc version of the board prior to its formal establishment. Paul Purser's notes from that initial meeting on June 9, 1966, included comments about IESD being "handicapped in their camera development work" because of insufficient communications on camera requirements. More specifically, Purser suggested that, "IESD and FCSD should develop closer working relations and interchange of information on flight crew camera requirements." [62] Whether or not this conflict early in communications resolved itself, it appears the decision at MSC was that IESD manage "the development and provisioning of all photographic instrumentation," per the final version of the Camera Development Review Board's charter from September 8, 1967. Despite any actual control of the research, development, and procurement held by IESD, documents from well into 1968 from Warren North, chief of FCSD, to the Systems Engineering branch of IESD responsible for camera work indicate the strong influence of FCSD's preferences. North listed specific pieces of equipment and technical specifications for parts of the Hasselblad system needed for lunar photography from the command module, strongly recommending their purchase for greater cost effectiveness and weight savings.[63]

Hasselblads were far from the only piece of camera equipment on board Apollo spacecraft. Photographic training and mission time included the competing interests of still, movie, and television imaging. The ability of television camera manufacturers like RCA and Westinghouse to miniaturize studio camera technology for use inside and outside space vehicles is truly one of the greatest technological achievements of this era. NASA prioritized this work because of the expanding medium's immediacy for public relations.[64] And while working with television cameras added a layer of complexity to mission timelines,[65] still photography moved forward with relatively little difficulty. Kodak provided a thin-base film to maximize the number of frames available in each film magazine, reducing the number of magazines needed during flight. Flight Crew Support staff such as Jeff Bremer recalled this as one of the most advantageous elements of Apollo still photography. It maximized the exposures per roll, but made for tricky processing by the photo lab.[66] In fact, the PTL's Brinkmann pushed for a number of technical elements that would maximize the amount of photographic information, for the scientific, technical, and public audiences, requiring prioritization of selecting specialized cameras and films.[67] He stated, "the PTL *recommends that the Apollo Spacecraft Program place a high priority on securing a maximum volume of photography.* Film is light in weight per unit (about 1 gram) and has a great scientific impact per unit" (italics appear as underlined in original).[68] From the professional, image-processing side of the house at NASA, there was strong support for high-quality equipment and adequate time dedicated to the work in order to make a significant impact with images.

According to Bremer, a technician responsible for procurement, flight-testing, and qualification for each Hasselblad, the relationship between NASA and Hasselblad was mutually beneficial. After some issues with contractors during Gemini for the film magazines, Paillard and Hasselblad stepped in during Apollo to provide NASA with technical support. Bremer recalls that:

> We had a wonderful relationship. At that time, NASA was writing fixed price contracts, no incentives, no extra profits. We never had any cost overruns. We learned that Victor Hasselblad, the founder and owner of the company, had a great interest in the U.S. space program and he committed any and all of his resources to make our equipment. He set up separate assembly lines and parts bins and inspectors for our equipment.[69]

At least in preparing cameras for missions, the design, manufacture, and preparation appears to have been quite smooth. Once in space, however, it came down to the training of astronauts and their ability to overcome difficulties to capture desired images that determined the ultimate success or failure of using this single manufacturer's cameras for all still photography. With Hasselblad reconfirmed as the camera of choice for the Apollo program and on the lunar surface in a slightly modified form, astronauts could begin, or, in the case of astronauts who flew during Gemini, continue to train for photographic work during their spaceflights.

As expected, with three crewmembers on Apollo, additional technical elements to the mission, and more flight time, the need for photographic documentation also increased. In August 1967, for the first time in the U.S. space program, Deke Slayton made a formal request for astronaut training in photography. A memo from Slayton to an administrator in charge of contracts and procurements requested that a formal request for proposal, "be issued for the development and presentation of an astronaut general photography training course," outlined in an attached document.[70] The request continued to outline primary and secondary purposes for the course. These span from the priorities of learning photographic basics and making in-flight decisions about photographic situations to the less important needs of learning the effects of radiation on films, preparation of a self-guide on photography, and giving the flight crew support staff the appropriate information on how to continue preparing astronauts for photographic tasks.

Though Slayton's request appears out of thin air because it referenced no preceding events, requests, or studies, it is possible that a study completed a few weeks earlier prompted his action. At a conference regarding lunar science and exploration directed by Wilmot N. Hess, the Apollo program director for Science and Applications, summarized one element of the study in Appendix B of the report, which was the importance of adequate photographic training for astronauts. A subset of scientists and photographic experts, including lunar geologist and later astronaut Harrison Schmitt, Polaroid camera inventor Edwin Land, geologist Eugene Shoemaker, and astrophysicist Thomas Gold, recommended techniques, technology, and training for lunar photography. Specifically, the panel suggested that:

> potential astronauts for the Apollo lunar landing missions be especially trained in photography, generally, and in particular, in the use of the Hasselblad camera for early Apollo missions. The astronauts should be given Hasselblad cameras with viewfinders similar to those to be used on the Moon. They should be encouraged to take many pictures of their everyday activities and surroundings. These pictures should be extensively criticized by an expert who would give each participant constructive advice. Such a program will help instill an instinct for good documentary pictures on the lunar surface.[71]

Follow-up subpanel meetings through late 1967 and early 1968 reflected ongoing discussion of increased astronaut photographic work to satisfy the needs of scientists and the public.

Based on the timeline established in Slayton's request, the possible three-day course consisting of approximately twenty lecture hours would educate astronauts by fall 1967, and provide self-study materials for those brought into the program later. The statement of work identifies Helmet Kuehnel as the project manager with Jeff Bremer and R. Thompson of the FCSD as the technical monitors for the contract.[72] In my interview with Bremer in 2012, he confirmed the utility of the course as it was given, stating that, "The crew response to my setup of classes was terrific. They really like[d] the way we presented the way to adjust the exposure settings for each type of picture, i.e., desert, ocean, cities, farmland, etc."[73] By all accounts, primarily from the astronauts themselves, handling photographic equipment posed no great difficulty to them, and they felt well trained for that element of mission work. Apollo 15 Command Module Pilot (CMP) Al Worden, selected in April 1966, described his transformation from complete beginner to proficient photographer:

> I first became interested in photography when I was the backup Command Module Pilot (CMP) for Apollo 12. . . . As [Dick Gordon's] backup I needed to become familiar with his flight plan and the experiments he was to perform. That meant that I should become familiar with the camera and the photography called for in the flight plan. I discovered that as I became comfortable with the camera and the photos I took, I started to enjoy the process of taking photos and improving my technique with time . . . , but because it was so enjoyable and satisfying, I worked very hard on my own to become proficient.[74]

Though Worden did not recall the training course designed by Bremer and the FCSD staff, he expressed no concern over the support from NASA in his effort to prepare for mission photography for Apollo 15. In fact, by the time of his flight an entirely new photographic project was added to his list of work: detailed lunar observations via his handheld camera in conjunction with a large-format metric camera and panoramic camera installed inside the service module (SM) of the spacecraft.[75] Worden became the first CMP to make a translunar extravehicular spacewalk to retrieve those rolls of film, later studied by geographers and scientists to create detailed maps of the lunar surface.

As with virtually every process and procedure planned for space missions, the steps not only required hours of training and time in simulators for preflight practice, but also a set of printed documents for in-flight use. Procedure manuals offered astronauts a written set of checklists and reminders for each step of the mission, making spaceflight more like scripted acting than free choice. Operating cameras and other equipment became part of the same scripts, or in some cases, required entirely separate documents due to the sheer volume of work. For each Apollo mission, a photographic plan complete with predetermined camera settings, expected mission elapsed times, and target listings prepared astronauts prior to flights. For the flight documentation carried in the spacecraft, those plans and technical details were distilled down into small notations in the official flight timelines embedded in portable flight plans and checklists (Figure 1.17).

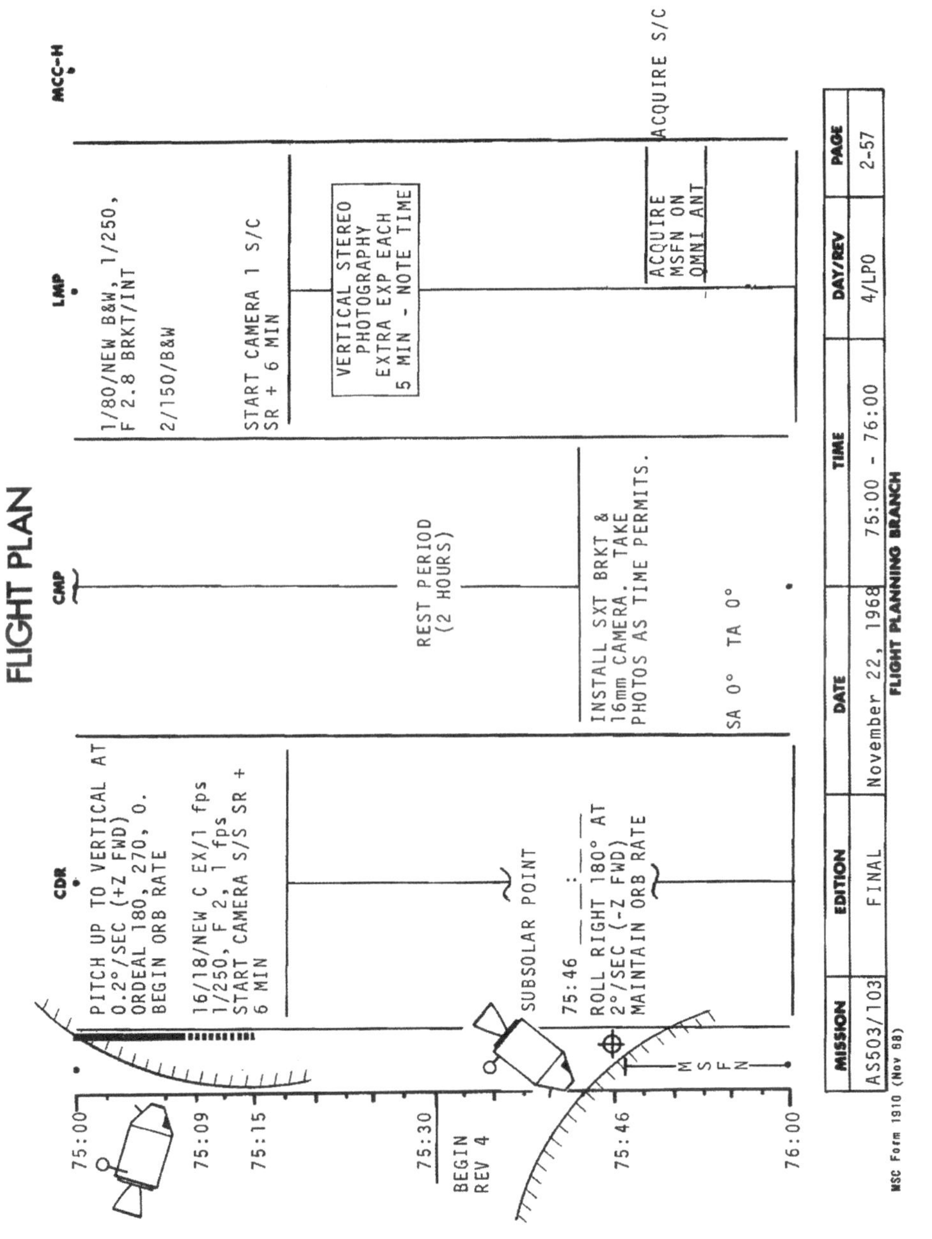

Figure 1.17. Apollo 8 flight plan page for start of fourth orbit, *Earthrise* image captured at approximately MET 75:47 (NASA).

Astronauts, however, with their independent and motivated personalities, learned enough from their photographic training to avoid sticking strictly to these mission scripts when they found an opportunity to accommodate photography or felt some visual inspiration. Within moments of putting his boot prints onto the lunar surface, Neil Armstrong, known for his deliberate and cautious work as an astronaut, snapped numerous photographs with his Hasselblad Electronic Data Camera before collecting the single contingency sample required in the mission plan as his first lunar activity.[76] Scolded by Mission Control and Lunar Module Pilot (LMP) Aldrin for not proceeding first to the sample, Armstrong wrapped up his photography, saying later, "It was going to take somewhat more effort to get that sample—to get the equipment and the container for that sample—than it was to get a few pictures."[77] In that incredibly historic moment, not only was there a logical reason for changing the order of activities because the scoop equipment took more effort to assemble and the camera was already in hand, but Armstrong, as historian James Hansen puts it, "was so intent on taking a few pictures," because capturing those sights was important too.[78]

More often repeated is the story of veering off the flight plan for photographic opportunities on Apollo 8, when, on the fourth lunar orbit, the three astronauts saw the Earth emerge from behind the Moon for the first time. Numerous sources cite this as an unscripted, unplanned moment of photographic activity, with Frank Borman, Jim Lovell, and Bill Anders scrambling to find cameras and magazines. While the exact moment may have caught them off guard, evidenced in the surprised tone of their voices in the mission recording, scientists and mission planners long expected the *Earthrise* moment to be captured during Apollo 8.[79] A notation for the image even appears in the final mission photographic plan. But, at that exact moment in their mission, the page from their flight plan shows that the only planned photography was an ongoing sequence of images captured automatically during their lunar far side pass. Strangely, this crew-selected target hardly received mention in the post-mission analysis documents produced by NASA, including the official mission report and a special publication on photography on Apollo 8.[80] For the three astronauts and people around the world, the sight of Earth from lunar orbit would come to symbolize the sublimity of human spaceflight.

Astronauts took sometimes-extraordinary measures to capture images they themselves planned. During Apollo 12, astronauts Pete Conrad and Alan Bean contrived a way to take a photo of themselves, together, during their time at

Surveyor 3. Bean asked flight crew equipment managers to make a timer for the Hasselblad he and Conrad carried so that they could set up the camera temporarily on a solid surface and then walk some distance away to get themselves both in the frame. This unplanned photo, which the astronauts were certain would not receive approval as part of their schedule, meant Bean had to hide the timer inside a tool carrier bag and fish it out later while on the Moon. Unfortunately for Bean, Conrad, and audiences back on Earth, the timer was hidden too well, and he never located it in time to keep the mission on schedule. Bean later reproduced how he thought that photograph and the experience would have looked in some of his own paintings.[81]

On Apollo 16, astronauts John Young and Charlie Duke executed a series of photographs, captured on television and Duke's Hasselblad over five frames, to show Young leaping off the lunar surface and saluting the U.S. flag planted nearby. Without a viewfinder on the camera, the astronauts repeated the leap photograph to ensure they captured it, having recorded it as moving footage as well on the remotely controlled television camera on the lunar rover. Duke described a moment during the mission captured perfectly on the first attempt: "Hey, John, this is perfect, with the LM and the Rover and you and Stone Mountain. And the old flag. Come on out here and give me a salute. Big Navy salute."[82] What was so difficult for Young about the leap salute was getting his body weight plus that of the space suit and his personal life support system (65 lbs in 1/6 gravity, just under 400 lbs on Earth) off the ground while and maintaining his balance. Commentary by Duke and *Apollo Lunar Surface Journal* author Eric Jones attributes the success of it to Young's comfort level with the environment so soon into the mission (just an hour into the first moonwalk) and his superior balance and coordination.[83]

Apollo photographic work was not free of its problems, though, and malfunctions, the use of incorrect settings, or the wrong film magazine became fodder for technical debriefings held after every mission. From the memorandum generated from the Apollo 10 photographic debriefing, it is clear that astronauts were "not quite happy with the photographic equipment."[84] Not only did they experience what they called persistent jamming of the film, but the cameras also malfunctioned, the large 250mm lenses took up too much space and were unwieldy, and they needed to swap one viewfinder between cameras too often. In the weeks leading up to the monumental Apollo 11 landing, concerns were raised over having the right equipment to correct for the effect of ultraviolet light on color film used in the Hasselblad cameras.[85]

Despite problems with equipment, and recommendations for remedying that for future missions, the crew commented that they felt well prepared for photographic work, especially with the photographic plan in place and available during the mission.[86] Upon return of flight hardware, NASA technicians and engineers such as Jeff Bremer and James Ragan took steps to repair the cameras, modify future flight hardware according to recommendations by prior crew experiences like that of Apollo 10, and update photographic plans to account for operational recommendations or additional desirable scientific and technical work. Bremer recalls that, "After we received the flight hardware we had to run a Post Flight acceptance test to prove that the camera equipment still functioned properly after the flight. It was basically shoot film with each lens, magazine, and camera, and evaluate it. Once the equipment passed the test, it was put back in bonded storage for use on later flights."[87] Nowhere in these reports is the source of any problems with photographic equipment really identified: equipment flaws or user error. Some instances such as using the wrong film magazine with the wrong settings or using the wrong developing technique in the photo lab did occur.[88]

The Apollo program provided astronauts with professional tools for handheld photography and the training needed to accomplish the goals of many audiences. With a greater focus on science in the J-series missions of Apollo 15, 16, and 17 in late 1971 and 1972, specialized equipment for astronomical and geological work accompanied the Hasselblads. Apollo 17 astronauts even had a Nikon 35mm camera on board for more personal, interior photographs that had more public appeal than scientific value. Though the program would end, technological experimentation continued into Skylab and Apollo Soyuz Test Project, preparing astronauts for longer duration orbital missions on a Space Shuttle.

The selection of camera technology, like equipment used for spacecraft and launch vehicles, was an iterative process through the Mercury, Gemini, and Apollo programs. From ad hoc modifications of store-bought automatic 35mm cameras to purchasing the premodified, professional quality, medium format cameras for lunar exploration, NASA learned gradually what technology would suit a variety of audience needs without compromising the quality, flexibility, and reliability required for unpredictable scenarios presented by human space travel. The process matured when NASA managers bought into the arguments made by engineers, that what the astronauts were seeing and doing presented a unique opportunity to create a visual narrative, and that a human behind the lens gave audiences an understanding of spaceflight by proxy.

Professional tools helped astronauts carry out their duties as surrogates for scientists, explorers, artists, poets, historians, and photographers back on Earth. NASA sought out an effective system that would go beyond the expectations of professional photographers and function flawlessly in the harsh environment of space. Despite the clear ease of use with 35mm cameras, an option only provided at the beginning and end of this period, few could deny that the image quality and potential information to be gathered from imagery (e.g., photogrammetry) created with a medium format 70mm camera could be surpassed. To fly to the Moon with a camera anyone on the street might have almost seems underwhelming and maybe unprofessional. The expectations from different constituent groups required, at least initially, gathering images that satisfied all needs, with professional equipment not just appeasing astronauts who likely felt more comfortable with simple camera equipment. The astronauts carried out photography as part of their duties, with well-defined procedures laid out in training sessions, pre-mission documentation, manuals, and notations integrated into the official flight plans. This level of careful planning for all aspects of spaceflight work gave astronauts a streamlined plan for carrying out a mission.

The significance of these cameras to the mission and to the legacy of spaceflight cannot be underestimated. They are particular technological artifacts that contribute to our understanding of human spaceflight even today. The ultimate purpose of the images created with them yielded documents of space exploration, while secondarily providing a visual rhetoric that would support additional funding for NASA work. Audience needs became the driver of image content and how NASA chose to distribute the catalog of astronaut photography. Photographs provided a substantial source of proof and reinforcement for the rhetoric developed by users of the images.

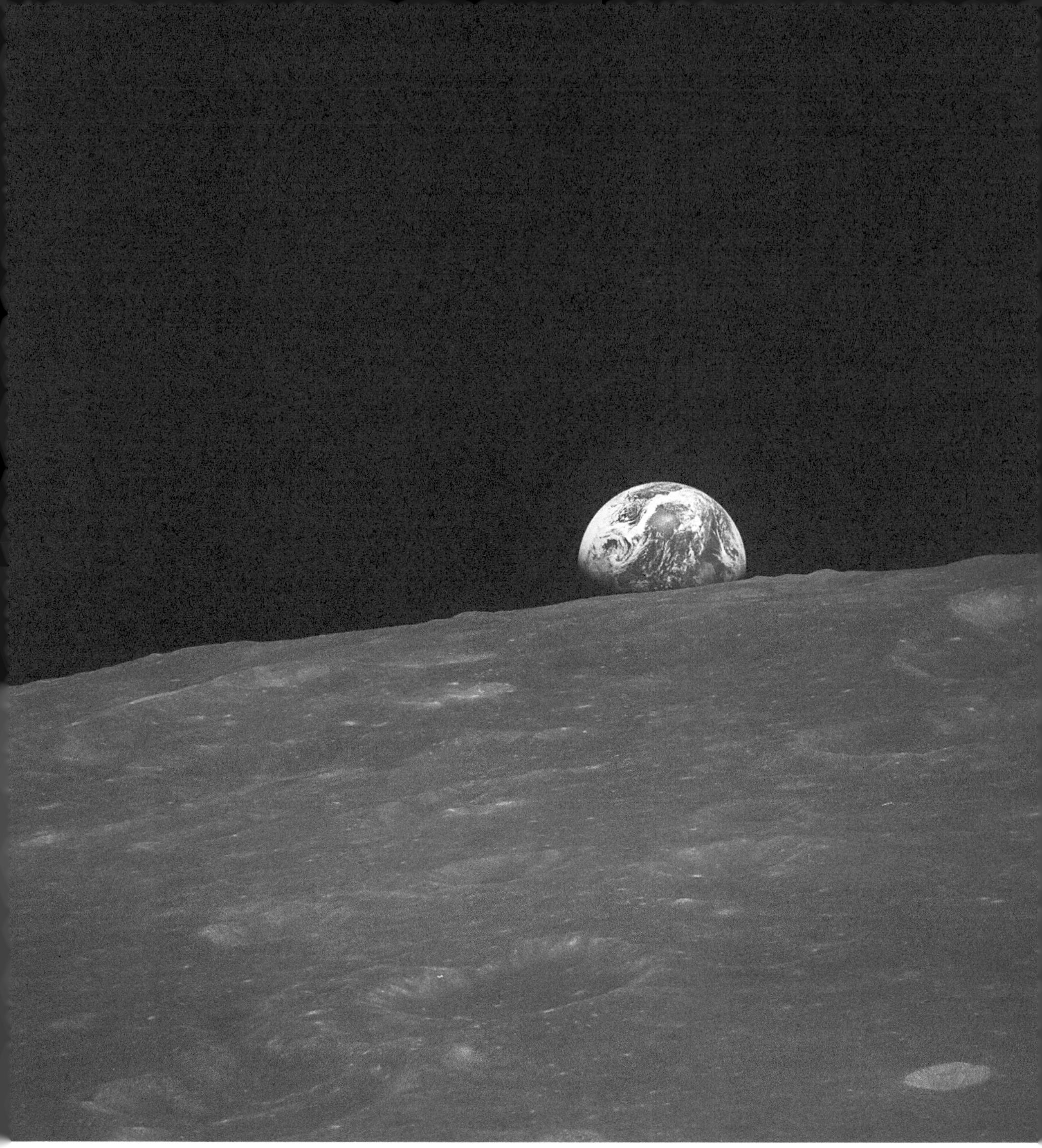

Figure 2.1. *Earthrise*, photographed by Frank Borman, December 24, 1968, AS08-13-2329 (NASA).

CHAPTER 2

Photographs for Every Audience

"But, the most impressive aspect of the flight was [when] we were in lunar orbit. We'd been going backwards and upside down, didn't really see Earth or the Sun, and when we rolled around and came around and saw the first Earth rise. [T]hat certainly was, by far, the most impressive thing. To see this very delicate, colorful orb which to me looked like a Christmas tree ornament coming up over this very stark, ugly lunar landscape really contrasted."[1]

—*William A. Anders, Lunar Module Pilot, Apollo 8*

In the wake of the Apollo 1 fire on January 27, 1967, NASA engineers and contractors worked to build a safe, operational Block II command module in order to resume the human spaceflight program. The first crewed Earth-orbital Apollo test mission, Apollo 7, flew in October 1968. For the first time since the brief, low-quality television broadcasts by Gordon Cooper during his Mercury *Faith 7* flight, a series of live television broadcasts by the astronauts brought the realities of spaceflight to televisions around the world. The success of Apollo 7 allowed NASA to commit to a lunar orbital mission two short months later, with plenty of cameras in tow.

The prior August, George Low, head of the Apollo Spacecraft Program Office, suggested a significant shift in the plan for Apollo 8. Grumman Aerospace announced that a complete lunar module for Earth-orbit testing would not arrive at NASA until early 1969. That meant flying it on Apollo 8 as scheduled was impossible in late 1968. Low began considering a radical plan that would send the Apollo 8 spacecraft to lunar orbit, motivated largely by delays to the lunar module construction schedule and reported signs of Soviet progress toward a human mission to loop around the Moon. With President Kennedy's end-of-the-decade challenge of landing on the Moon looming, NASA put the entire Apollo program to the test by using a Saturn V to put the Apollo 8 spacecraft in orbit around Earth's only natural satellite.

After its December 21 launch and two-and-a-half-day trip to the Moon, on Christmas Eve 1968, Frank Borman, James Lovell, and William Anders fired their service module propulsion engine to place the vehicle in lunar orbit. Almost as soon as they began their second circuit, Anders and Lovell pulled out camera equipment and plunged into their busy photographic plan while Borman attended to spacecraft operations. Their work focused on mapping the lunar surface (especially the far side), detailed images of landmarks, and most importantly for planning, extensive photography of features near the targeted first landing location at *Mare Tranquillitatis* (Sea of Tranquility). The flight plan had technical camera specifications laid out to include f-stops and other settings to make camera use simple and almost "point and shoot," despite the crew not having automatic cameras.[2] This provided more time to focus on targets and less on finding settings for specific conditions. Even preparing for situations such as dim light photography, best done with special film, received attention ahead of time, and film magazines for such photographs had special labels and notations in the flight plan. Mission planners worked in advance with technicians, the crew, and equipment contractors to ensure that no guesswork remained when astronauts used camera equipment as they circled the Moon. Other than smudges on the windows due to the off-gassing of sealant used between the window and the frames, the crew had few hurdles to completing the task of capturing scheduled targets of opportunity as well as some spontaneous "crew selected" images during their flight.

During the first two orbits, the spacecraft followed an elliptical path in order to increase the safety of the first-ever lunar-orbit burn. The astronauts rode backward on approach to the Moon, the main engine pointing forward for firing it again to circularize their path at the beginning of the third orbit. Borman then rotated the vehicle so that the windows pointed at the Moon for a picture-taking sequence. The fourth orbit offered the astronauts a spectacular view as they came around to the Earth-facing side. Anders exclaimed, "Oh, my God! Look at that picture over there! Here's the Earth coming up. Wow, is that pretty!"[3] The crew scrambled to find the right film magazines for their Hasselblad cameras to capture the view now known as *Earthrise*.

Borman snapped the first photograph (Figure 2.1), though it was in black-and-white and closely resembled an image taken by the first Lunar Orbiter spacecraft two and a half years earlier (Figure 2.2). Less than a minute later, Anders swapped out a black-and-white film magazine for one with color film and captured the iconic *Earthrise* color image as part of a sequence of nearly twenty photographs with different lenses and from different angles. Since photographs cannot speak for themselves, it took the eyes of NASA's corps of professional photographers, public affairs officers, and other amazed viewers on Earth to launch that single frame from Apollo 8's Magazine B to the pages of publications around the world. Without any inherent scientific or technical value, the recognition of its uniqueness and the legacy of its use and reuse by magazines, newspapers, books, and the environmentalist movement cemented its place as one of the most iconic images of the twentieth century.

Through the landing of Apollo 11 in July 1969, NASA activities garnered intense and near-constant media attention. While that interest waned somewhat after the first successful landing, the provision of still photographs increased exponentially, allowing even greater focus on the needs of those most stymied by the lack of opportunity to explore in person: the scientific community. Between Project Mercury, when still photography began as an afterthought, and the last three Apollo missions, when astronauts mapped the Moon using special high-resolution cameras, the influence of NASA and non-NASA audiences significantly altered the trajectory of astronaut photographic work. Resources, training, and planning increased many times over just as mission profiles grew more complex.

To understand how NASA produced a comprehensive visual rhetoric using still photography requires understanding the intent of managers to serve as many needs as possible. Their decision-making process demonstrates the confines of performing scientific pursuits within the federal government bureaucracy. First, interested communities who required spaceflight visuals by astronauts needed the ability to shape mission photographic plans. Who were these audiences, how did they submit their requests, and what were their expectations in regard to visual representations of spaceflight? Second was the prioritization of photography within mission plans. How did NASA manage requests for images and integrate those into the guidance

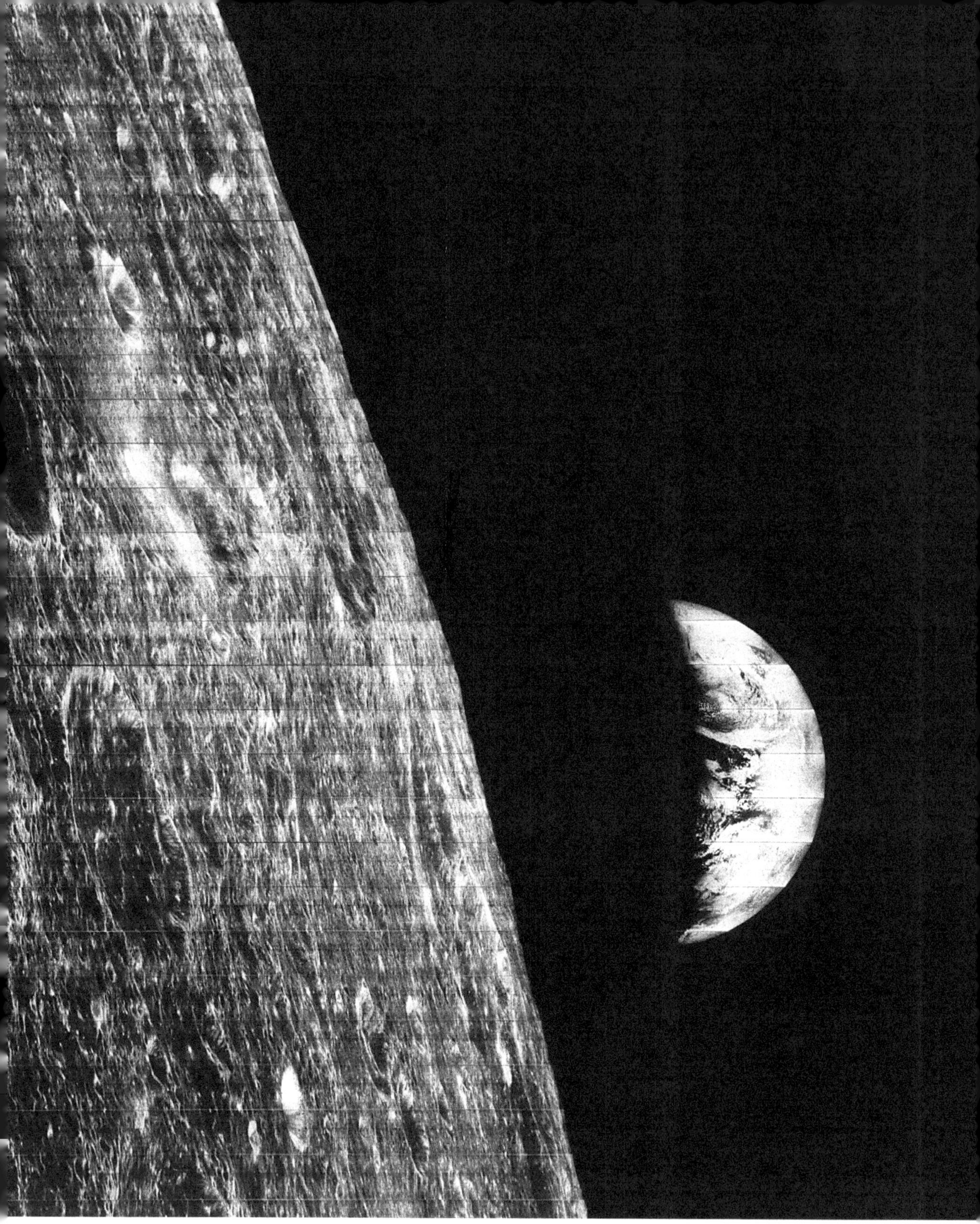

Figure 2.2. *Earthrise*, photographed by Lunar Orbiter 1, August 23, 1966, Frame 102-H2 (NASA).

given to astronauts before and during flights? Third, when film magazines returned from missions, astronauts left their photographs in the hands of technicians and other NASA personnel who distributed those films to audiences. Who physically managed the processing of flight films and how did they transfer these images to scientists, engineers, and public affairs officials for final disposition? Putting additional hands on the photographs altered how audiences interpreted and understood the experience of early human spaceflight. Answering these questions explains how still photography reinforced NASA's space exploration rhetoric in the 1960s and 1970s.[4]

IMAGE AUDIENCES

How managers and public affairs officials selected images for publication may seem somewhat self-evident. If the amazing photographs seen in newspapers and magazines were the only ones returned from space, the job would be simple. Astronaut photography instead provided hundreds of images during Mercury and Gemini and nearly 20,000 from Apollo. As with previous expeditions on Earth, images returned from astronaut exploration needed to fit a preconceived notion of the project's meaning in order to perpetuate support. As documentary, scientific, and technical data, images provided information while also supporting the sometimes-fanciful conceptions of exploration held by the audiences they served. The government required visuals to justify their support and public audiences needed that same information to understand spaceflight. Scientists and engineers needed evidence to support continued research and technical work so images could rarely serve a single purpose. The ultimate legacy of NASA photography is how certain photographs were so easily popularized and their use in unexpected places. Identifying the audiences themselves and the means by which they received the images will give some definition to the types of images returned by astronauts.

NASA itself was the primary audience for space photography provided by satellites and astronauts during orbital and lunar missions. From top to bottom, NASA audiences required visual materials to complete their missions in administration, engineering assessment, scientific research, mission planning, and the communication of accomplishments to audiences outside the agency. The vast majority of the audience, however, fell outside the agency, so images needed to bridge a gap between the specialists and the less informed by having a broad appeal. Still photography, movies, and television broadcasts fell far behind mission-critical requirements for vehicle operation and safety, but they became the primary means of connecting real spaceflight to those familiar with fictional space travel from television and movies.

Early NASA planning and management of astronaut schedules put emphasis, quite understandably, on operating and assessing the operation of spacecraft. As the number of orbits increased, so did the expectations for how much work to schedule. In a preliminary flight planning document for John Glenn's 1962 orbital flight, Helmut Kuehnel—head of what was then called the Flight Activities Section—suggested to management that the focus should be on simulation leading to a formal plan for mission control, systems evaluation, and observation.[5] Kuehnel emphasized the first two elements as critical, and though he included the reference to observations, he made no specific mention of photography. He instead suggested the recording of observations in "voice recording and written notes and sketches."[6] His omission of photography lends weight to Glenn's story of a late addition of a camera to his flight equipment. Examination of Mercury flight plans and those through the end of Apollo confirm the secondary nature of observational work by astronauts who naturally treated spacecraft operation and crew safety as their primary responsibilities.[7] Not until Gordon Cooper's *Faith 7* in 1963 flight did a final plan call out astronaut activities not related to spacecraft operation or safety in a prioritized order, placing photographic experiments at the third, fourth, and seventh positions.[8] While astronaut photography never superseded critical mission elements like operations and safety, the importance of visual representations increased quickly through Project Mercury and into Project Gemini, when the addition of another astronaut to the flight made more mission time available for such work.

In a very visible way, Gemini mission plans call out the audiences for photographic experiments carried out during flights. While experiment designations continued in Apollo mission documents, photography no longer appeared described as a series of experiments, but as part of the regular procedures carried out by crewmembers.[9] Mission preparations included a thorough photographic plan that broke photography down into either operational or scientific work. Mission planners then included photographic plan elements within the official flight plan timeline and concluded the process with formal photographic debriefing sessions following the flight. Management summarized their photographic philosophy for Apollo in the *Apollo 8 Photographic Plan*: "The objectives in this plan that aid the assessment of this and future missions and the planning of future missions are given first priority and are listed as operational objectives." While other photography was desirable, the plan states, targets of opportunity did not displace other critical mission operations.[10] While NASA rhetoric called for visuals of spaceflight to enlighten a broad audience, internal communications and documentation supports the idea that insiders worked to put the agency's spaceflight goals, engineering needs, and scientific investigations above other concerns.

By sheer numbers, one might surmise that science was the driving force behind the images taken by Apollo astronauts. Nearly 18,000 individual usable frames of film resulted from Apollo astronaut work, building upon the over 2,600 frames taken during Gemini, and the paltry 332 frames from Mercury. The spacecraft, in essence, performs the role of the laboratory as Peter Galison describes the place in which images of scientific activity are created.[11] Moreover, while documentation and operations show science was only a side benefit to spaceflight in the eyes of NASA managers, the yield of photographic material offered a storehouse of images for ongoing scientific research. With a voice in photographic planning and resource allocation, giving them ability to steer the amount of mission time allocated to photography, the scientific community of planetary geologists and astronomers shaped the targets of those images. Image content in this case largely excluded the human explorers in space or on the Moon in favor of scientific and engineering targets. As percentages of total images, those with an implied or obvious scientific purpose vastly outweigh those featuring people by nearly 20 to 1 (Table 2.1). In terms of volume, what these figures indicate is a clear distinction between who the audience was and was not. The public needed familiar human forms to interpret images, but scientists looked for diverse content types supplied in plenty through astronaut photography.

The story told by this table, on the face of it, is that people appearing in images—what would most easily connect viewers to the astronaut experience—made little difference to NASA in determining photographic targets. Alternately, it is possible that planners and engineers writing up mission plans never discussed representing humans and their activities through still images. Only a handful of the images with human figures in them appear staged or posed, with most being clearly accidental or of boot prints or shadows. While NASA staff photographers

Table 2.1. Tabulation of all images featuring humans (including shadows and footprints).[12]

Mission	Images taken	Images featuring people[13]	Percentage of total
Friendship 7[14]	75	0	0
Aurora 7	206	0	0
Sigma 7	14	0	0
Faith 7	37	0	0
Gemini 3	25	0	0
Gemini IV	336	29	9%
Gemini V	349	3	1%
Gemini VII	426	0	0%
Gemini VIA	198	0	0%
Gemini VIII	18	0	0%
Gemini IXA	348	16	5%
Gemini X	354	7	2%
Gemini XI	231	12	5%
Gemini XII[15]	398	41	10%
Apollo 7	531	12	2%
Apollo 8	865	0	0%
Apollo 9	1373	68	5%
Apollo 10	1436	5	.3%
Apollo 11	1408	77	5%
Apollo 12	2119	183	9%
Apollo 13	604	16	3%
Apollo 14	1342	85	6%
Apollo 15	2640	321	12%
Apollo 16[16]	2801	511	18%
Apollo 17[17]	3985	655	16%
TOTALS	**22119**	**2041**	**9%**

like Richard Underwood offered guidance on image composition, only a few photographs appear planned other than if the intent was to depict a certain feature of the environment or technology. Closer examination of missions with the most representation of humans (i.e., shadows, boot prints, recognizable portions of the human body) in Table 2 show that they are also some of the most high-profile missions in terms of accomplishments. Gemini IV included the first U.S. spacewalk in mid-1965. Gemini XII was the last of its program and provided a final opportunity to perfect procedures necessary for Apollo, but also carried a very outspoken astronaut, Buzz Aldrin, who was overtly focused on nurturing a public persona through photos. The last three Apollo missions were far longer in duration and scientific scope, meaning more photographs of their work would include signs of their presence. The significance of the human in accomplishing goals on these three missions perhaps encouraged an increased attention to taking photographs of themselves, though few were more than incidental in nature. Of course, there is also the possibility that these crews were just predisposed to photographing each other more because of mission requirements. While these observations do not suggest an intentional neglect in scheduling photographs that included people, there was no obvious effort to help the Earth-bound understand spaceflight through the appearance of familiar shapes such as human figures, footprints, or shadows. The engineering, scientific, and bureaucratic audiences, then, benefited most from the incredible catalog of photographs produced in this period.

Administrators

For the highest levels of NASA management, images meant a great deal in the legislative game of securing funding. In this context, the idea of NASA's work was as important as its actual work. Photography captured by astronauts provided materials for increasing the visibility of the agency among elected officials and the public. Available documentation from upper-level managers at NASA indicates that they were involved in photographic planning discussions. As early as December 1962, Robert Gilruth became a regular recipient of memos regarding astronaut-captured photographs and their value inside NASA.[18] In this regard, the head of the center responsible for astronaut activities and mission operations was at least aware of how much NASA staff valued astronaut photographs as such evidence came from few other sources at that time. Another memorandum from public affairs officer Brian Duff to NASA Administrator James Webb in late 1963 summarized photographic work of astronauts and the press pool through Project Mercury. Duff's outline included how those products were useful for outreach and internal engineering needs.[19] Worth noting is the origin of the document in public affairs, revealing the pathway of visual communications expected in this early stage of human spaceflight. Everything was potentially useful for increasing the profile of NASA programs, and even the publicly facing wing of the agency needed to know what was available.

Additional discussion between NASA departments, including upper-level managers, occurred prior to the first Gemini launch. In December 1964, the Office of Space Science and Applications (OSSA) at NASA Headquarters approved an astronomical photography experiment and transferred funds for the grant to MSC.[20] The memorandum regarding the project went through the office of Gilruth, but approval and direction to carry out the experiment from Headquarters indicated administrator-level awareness and involvement in shaping Gemini photographic work. This is not an indicator, however, that NASA managers believed astronaut photography could provide much support to the agency's mission. That shift came with new views that placed relatable subjects—the astronauts—in the photographic frame and context of space. Images of Ed White's spacewalk were that transition.

By the time of Gemini IV, astronaut photography appeared on the radar of the most important NASA managers working on human spaceflight, including George Low and Robert Gilruth. While they released astronaut-captured photographs from all of the Mercury missions, no flights until the Gemini IV mission depicted the reality of human life in Earth orbit.[21] To NASA managers, the value of images emanated from the ability to see humans and their technological creations in the context of space. Alternatively, Underwood directed attention to what was perhaps the most valuable aspect of astronaut photography in the long term: Earth imaging. The ultimate goal for the space agency, however, was not the glorification of the humans doing the work, but the work itself.

Leading up to Apollo, managers called for a number of planning groups and processes for ensuring that astronaut photography covered all of the needs of NASA and its constituents. In early 1966, the Associate Administrator for Manned Space Flight George Mueller and the Associate Administrator for Space Science and Applications Homer Newell agreed to form a committee to analyze needs for lunar photographic data.[22] Their intention was to deal with current and future analysis possibilities in order to manage the valuable assets expected from lunar missions. Formal confirmation of the need for such a committee came, however, from an even higher level. In late 1967, the President's Science Advisory Committee (PSAC) Space Panel appointed a subpanel to study and guide lunar photography.[23]

Their initial meeting in fall 1967 involved discussions of developing a stereographic close-up camera later manufactured by Kodak and flown on a number of Apollo flights. Later meetings included executive sessions in which Mueller himself reviewed plans for early missions, with details on the agenda such as electrical fuses used in the Hasselblad cameras, batteries for a spot meter, and the possibility of using color film in geological photographs.[24] The Advisory Committee on Lunar Surface Photography included notable faces from inside and outside NASA, including Dr. Edwin Land, Dr. Eugene Shoemaker, and Nobel Prize winner Dr. Edward Purcell. Along with the other participants, the inclusion of such distinguished professionals on a lunar photography committee is a testament to the seriousness with which NASA treated imaging on the Moon.

Other advocates for photography, in this case lunar imaging from the CM, also came from the upper tiers of NASA management. Apollo Program Director General Samuel Phillips wrote to Robert Gilruth in March 1968 to request a review of orbital photography for scientific, operational, and contingency purposes.[25] His memo in this case refers to a cancelled plan for a Lunar Mapping and Survey System, a version of which flew as a three-part mapping system during the last three Apollo missions. Responding to the concerns of those in the scientific, engineering, and planning communities at NASA, Phillips advocated for thoughtful consideration of logistical needs for such photography, and the potential of missed opportunities with a vehicle and astronaut in lunar orbit. Astronauts in fact became crucial to early lunar mapping with their "bootstrap photography" until the mapping system—a camera, altimeter, and panoramic camera—flew in the experiment bay of the Apollo 15, 16, and 17 service modules.

Images could concurrently be useful and meaningful to many different viewers within NASA. Other than a brief moment following Gemini IV when NASA managers identified a unique value of astronaut photography upon seeing Ed White on his spacewalk, the vast majority of their later planning, discussions, and photographic work did not include any discussions of including people in the photographs. In a practical sense, mission planners identified overlapping interests of groups inside NASA early enough to make it possible to capture images for more than one purpose with the fewest number of cameras and equipment.[26] Perhaps in this way, NASA failed to identify the visually unique nature of their activities: the ability to put people into space and return images of that for public consumption. The ability of Earth-bound people to understand an image and find meaning in it changed when relatable elements such as people appeared in the photographs. In this way, the underwhelming number of images of astronauts in space up through the end of the Apollo program contributed to a general misunderstanding of spaceflight's dangers, and perhaps distortion of the utility of putting humans into space. The human spaceflight program then was ill-served by its own visual products: without a human in the picture, photography by robotic surrogates was seen as something accomplished just as easily. Ultimately, administrators bore responsibility for not considering adequately how image content familiarized nonscientists and engineers with spaceflight.

Engineers

While documenting crew performance through telemetry and debriefings served to provide details about missions, photography became an invaluable tool for documenting spacecraft and component performance as NASA gathered more flight experience. Early expectations by managers were that the data acquisition cameras, the small 16mm movie cameras used through this period, were adequate to capture photographic evidence of intricate technical procedures such as spacecraft rendezvous, docking, and lunar landings[27] (Figure 2.3). That would leave still photography for documenting single noteworthy moments relevant to the evaluation of mission operations. Early Gemini flight plans included still photography as part of experiments with more or less scientific intentions, but those documents make no mention of potential uses of images by engineers. In an interesting commentary on preparation for flight and using images for operational purposes, Gordon Cooper and Pete Conrad detailed their concerns about visual orientation during their Gemini V mission debriefing.[28] The mission, already loaded with photographic work, apparently did not include any visual references for Cooper and Conrad to orient the spacecraft in relation to the Earth's horizon, with Cooper complaining characteristically that "Never once did we have any darn thing to show us what out the window should look like."[29] Cooper went on to suggest that hand-drawn sketches placed in simulator windows would have helped, and encouraged planners to include fuel on the long-duration flight of Gemini VII so that Frank Borman and Jim Lovell could capture photographs to aid later crews in training for what they would see.

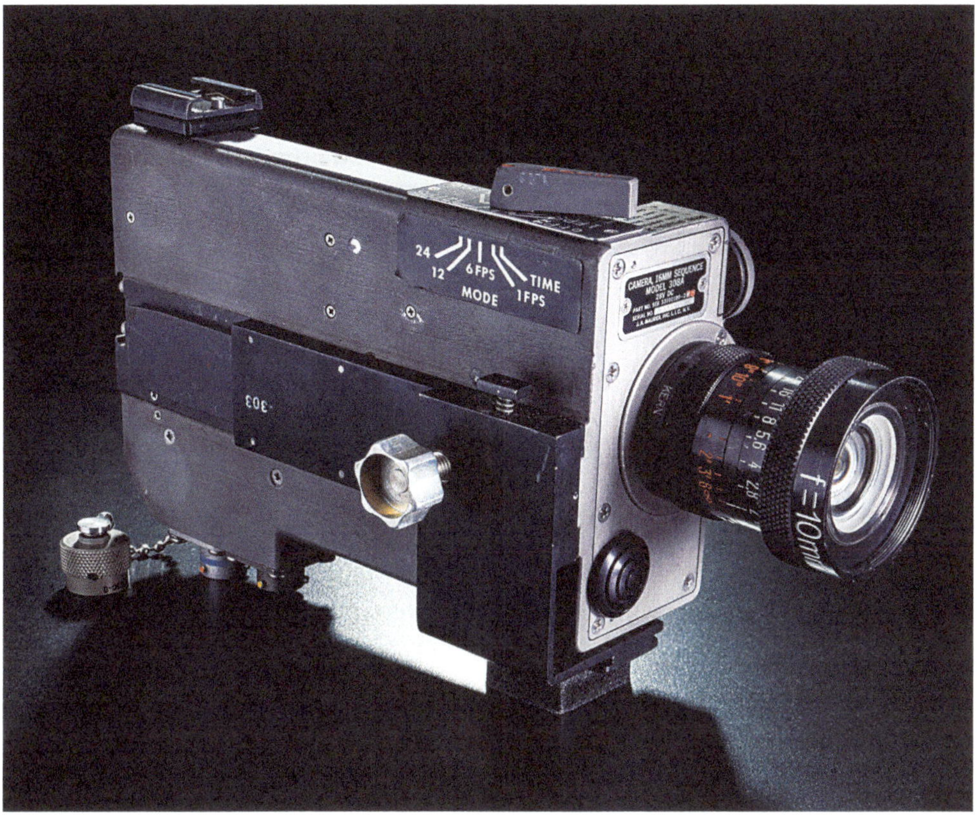

Figure 2.3. Data Acquisition Camera from the Apollo 11 Lunar Module, NASM Cat. #A20190311000 (Smithsonian National Air and Space Museum).

As activity increased in preparation for the Apollo lunar landings, so did the concern over documenting necessary stages of those missions with photography. In early 1968, Apollo program manager George Low detailed for MSC Director Gilruth's assistant Paul Purser the photography he envisioned at the time. They included, first, documentation of the lunar module at the landing site, followed by images of crew performance and local geography.[30] Still, as the Apollo program went on, suggested documentation of most engineering-related issues referenced the 16mm data acquisition camera, with passing references to useful shots possible with the Hasselblad cameras, such as the tire marks left in the lunar soil by the lunar rover.[31] Generally, however, engineering photography tended to require moving images and not still, indicating the engineering value of seeing processes that changed over time and not just single moments captured in photographs.

Examples from two particular events illuminate the additional value offered to engineers in still photographs. During the Gemini IX-A mission, the target vehicle did not properly shed its payload fairing, prompting the term "angry alligator" for its appearance (Figure 2.4). Engineers analyzed this and other photographs captured by astronauts Thomas Stafford and Gene Cernan to determine the cause of the malfunction. Visual evidence complemented the voice descriptions recorded on board the spacecraft and in transmissions to the ground, and this combination provided engineers at NASA the material they needed to prevent such mishaps in the future. Documentary photography also provided a level of accountability within NASA, never seen more clearly than in the frames captured by Apollo 13 astronauts of their damaged service module after jettison and prior to reentry into Earth's atmosphere. Unable to inspect the damage physically, the crew photographed the module as it tumbled away, giving NASA engineers their only visual clues as to what caused the near-catastrophic damage and cancelled the lunar landing (Figure 2.5).

Broadly speaking, engineers required still and moving images to document the ways in which their equipment and written procedures were successful or not in the space environment. With a visual record of a mission that yielded the greatest benefits to the most audiences, engineers helped establish image priorities that included documentation of operations, capturing evidence for future mission planning, determining when imagery could replace work, and recording scientific or other data.[32] Though dealing with human characters in one of the greatest explorations ever undertaken by humans, NASA engineers never acknowledged the centrality of people to the story of spaceflight.

Scientists

As mentioned, observational activities such as photography took a backseat to spacecraft operations in mission plans. Preparing astronauts for that work required input from interested audiences who needed visual materials. By the time of Wally Schirra's Mercury flight, the list of requested photographs shifted to specific descriptions in final flight plans, as opposed to random images captured out the window. Of the three astronaut-involved experiments listed in the Mercury *Sigma 7* flight plan, three involved photography and none rated higher than work associated with spacecraft control, medical observations, radio checks,

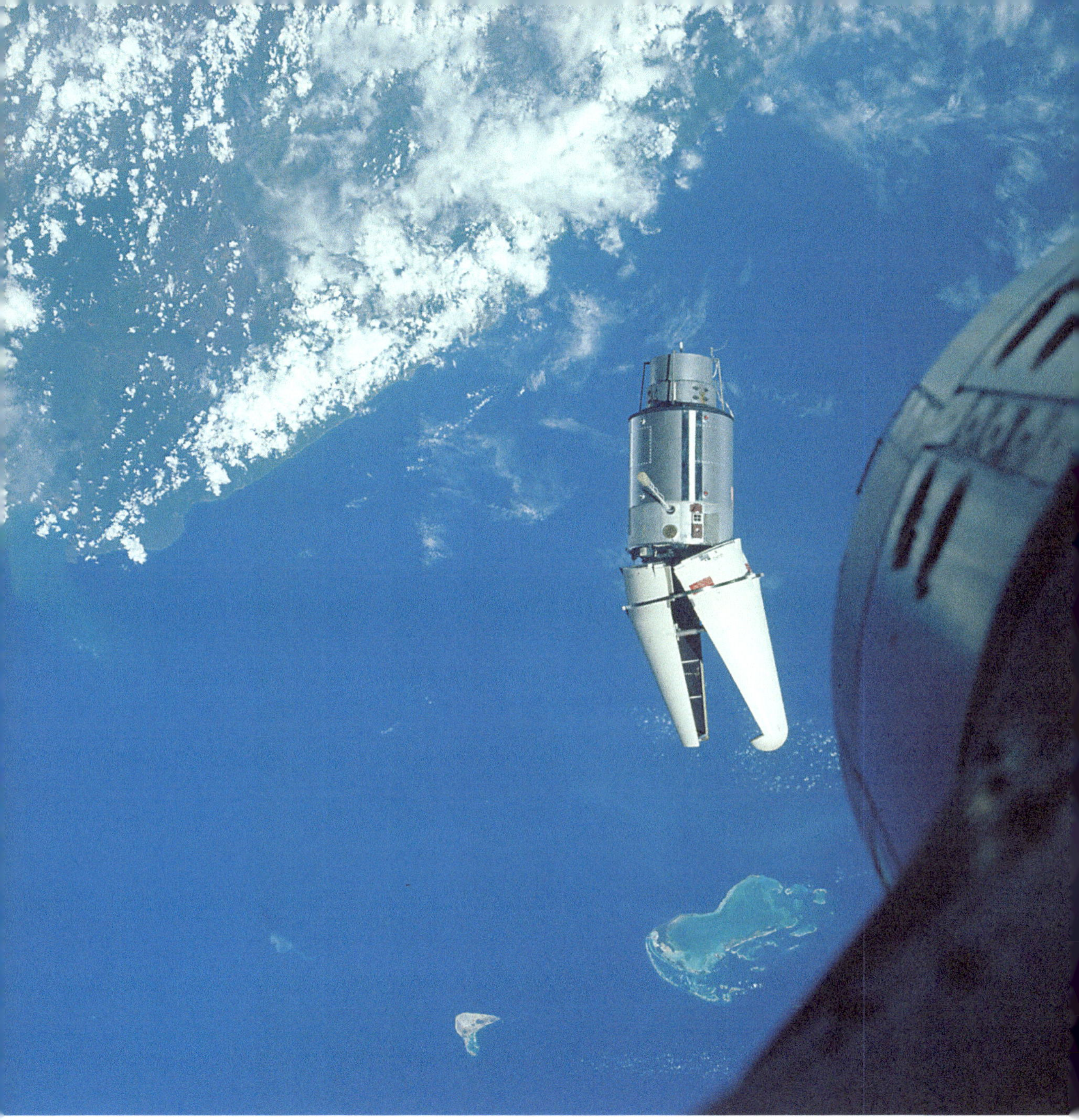

Figure 2.4. The Augmented Target Docking Adapter (ATDA) as seen from the Gemini 9 spacecraft during one of their three rendezvous in space, June 3, 1966, S66-37923 (NASA).

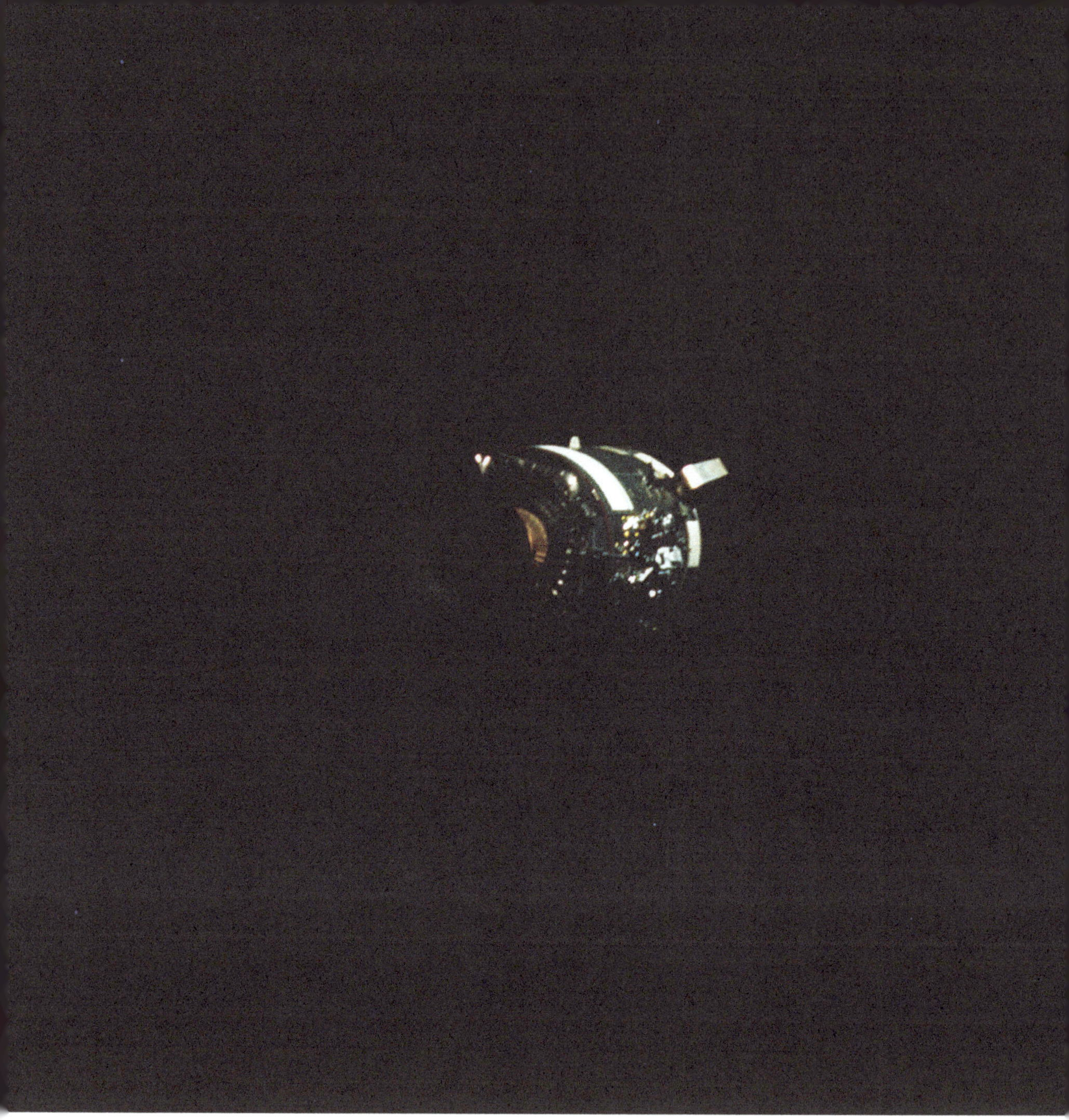

Figure 2.5. Magnification of the service module gave engineers their only look at the damage to Apollo 13, April 17, 1970, AS13-58-8459 (NASA).

and other scheduled operations.[33] Schirra later commented that his time in orbit was quite constrained, and he managed to capture only a few relevant photos to satisfy a request by NASA scientists for terrestrial features that could inform Earth resources work and possible mission planning.[34] In closing out Project Mercury, Gordon Cooper experienced an extensive post-flight debriefing that included very specific questions regarding exactly how he captured photographs for each of the science photography experiments.[35] The questions regarding equipment procedures, the process of sightings based on stars, and terminology used in discussing weather and Earth features reflected extensive preparation for these experiments. This indicates the future track for astronaut photography with scientific purposes.

Formalization of scientific photography during Project Gemini lead to greater description of the objectives, scope, and procedures in each mission's flight plan, hinting at extensive preparation by scientists to get it included on a flight. On Gemini IV, for example, the flight plan and an associated working paper provides details regarding a number of planned DoD, MSC, and scientific experiments involving cameras.[36] While two DoD experiments on basic object and surface photography were cancelled prior to the mission, one MSC experiment to capture two-color Earth limb photographs and two science experiments for synoptic terrain and weather photography did occur. The working paper provided top-down information on the experiments and the process for carrying them out. The final flight plan distilled the working paper information into systematic instructions along mission elapsed time marks and other procedural notations. Inclusion of scientific work in a mission plan at any stage of its development did not mean a guaranteed place on the mission though. On Gemini XII, the mission plan for photography had a late modification by the FCSD's Warren North, who sent a memo to the Gemini Program Office to exclude particular Earth photography work because the spacecraft and crew lacked preparation.[37] The plans for photography and any experiments remained in flux even during the mission, with real-time events playing a role in final products.

The scale and scope of reporting of scientific photography expanded with Gemini IV, the first long-duration mission. They developed into symposia, experiment reports, and other postflight documentation that required the input of both the principle investigators and astronauts as coauthors. While the astronaut-photographer for individual frames cannot often be determined, their coauthorship of reports suggests either a respectful nod to their role in capturing the photographs or an actual involvement in analyzing the results. As engineers and test pilots; however, few proclaimed any serious interest in observational scientific work.[38]

In NASA Technical Note D-3982 entitled "Terrain Photography on the Gemini IV Mission: Preliminary Report," Paul Lowman of the Goddard Space Flight Center (GSFC) and the two mission astronauts are shown as authors of the report based on photography in the S-5 Synoptic Terrain Photography Experiment. The last page of the report, however, acknowledges the input of others at NASA who helped interpret the photographs captured by McDivitt and White. This may indicate that the geographical and scientific information in the report did not require the astronauts' contribution.[39] That does not negate astronaut input entirely, as experiential information and observational recollections were undoubtedly valuable in writing a narrative of the mission's photographic work, but their direct influence on the use and perception of photography from Gemini missions seems highly unlikely. Successive experiment reports compared results from the same experiments across missions, making only a casual nod toward the astronaut contribution.[40]

For the Apollo program, the input of scientists into photographic plans is apparent across the documents from mission preparations, as well as in the sheer number of photographs collected that lack a human or technological subject. Science informed Apollo photographic planning from the top, as seen with the heavy representation by scientists on the PSAC subpanel on lunar photography: astrophysicist Thomas Gold from Cornell University, physicist Edward Purcell from Harvard, and astronomer James Baker of Harvard, who cocreated the Baker-Nunn camera system.[41] Much of their pre-Apollo discussion revolved around the development of a stereoscopic camera to capture close-up images of lunar soil and rocks, conversations that went back to at least 1965 among NASA scientists on functional requirements for cameras[42] (Figure 2.6). The experiments program involved numerous principal investigators from outside research institutions. The goal became one of minimizing equipment and maximizing scientific output, eliminating or minimizing overlapping interests between agency scientists and principal investigators. The cases of the stereo camera work and a special four-Hasselblad camera system for multispectral photography show that NASA knew that although it might compromise the needs of external scientists by reusing equipment for operational purposes, the urgency of the program made the move necessary at times.[43] Satisfying the scientific community inside and outside NASA had its own set of bureaucratic requirements.

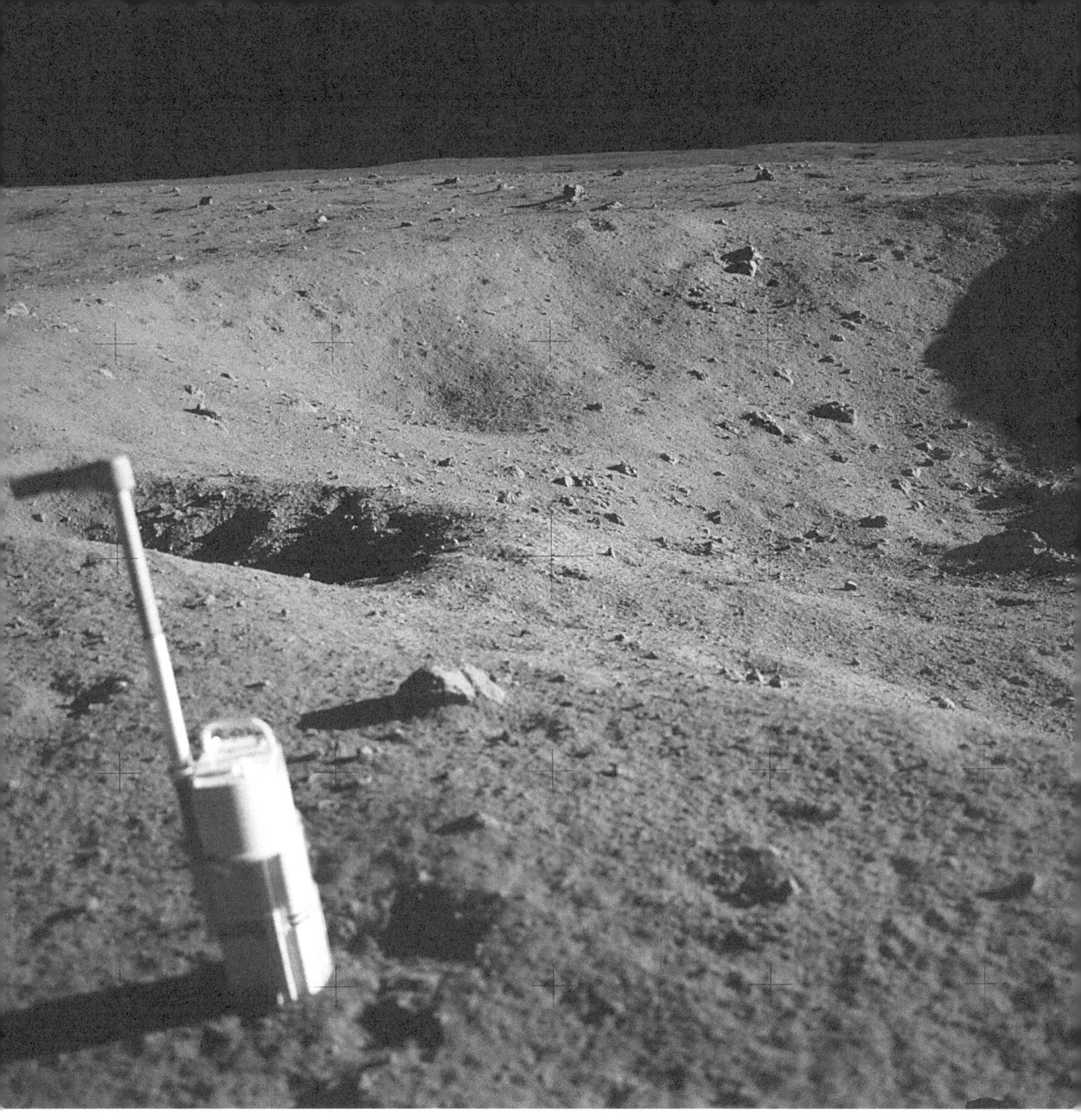

Figure 2.6. Apollo Lunar Stereo Closeup Camera during Apollo 11, AS11-40-5957 (NASA).

Calls for additional scientific photography went through the entire Apollo program, presumably based on the idea that when it finally ended, high-resolution and close-range photography of lunar features may not be possible for a long time. Still photography planned through the end of the program came to include an extensive set of Earth resources work done during the orbital missions, detailed imaging of scientific experiments set up on the lunar surface, and geographical mapping done on the last three missions.[44] Of special importance were singular opportunities to photograph phenomena from the perspective of astronauts traveling through translunar space. Preparations for the Apollo 13 mission included astronomical training to aid in photographing and describing Comet Bennett from the crew's unique point of view.[45] Photography of astronomical targets such as the L5 Lagrange point and the Gegenschein were also planned during lunar orbital activities. Apollo 15 astronaut Al Worden commented in an interview that he was most proud of capturing these due to their high scientific value.[46] As seen in Table 2.1, the overwhelming majority of astronaut photography had a scientific intent. That gave the motivated, goal-oriented astronauts a connection with scientific staff and the ability to satisfy specific needs, not just the abstract wishes of people like Richard Underwood with his objective for them to capture something meaningful. Though the handful of truly iconic images astronauts captured sustains our cultural memory now, the real substance of their work also continues to satisfy the needs of the largest community of users (i.e., scientists) of astronaut photography.

Other Government Agencies

NASA was hardly the only government agency in the 1960s looking at the value of human observations from space. The Weather Bureau, the predecessor of the National Weather Service, was a common customer when it came to requesting photographic work by astronauts, particularly during Project Mercury. To confirm camera lens filters needed for generations of Tiros and Nimbus weather satellites, the bureau requested a photographic experiment on Wally Schirra's *Sigma 7* flight to record different filter effects on a single film frame.[47] A letter regarding Schirra's photographic results and plans for more weather photography on Gordon Cooper's Mercury flight written by Robert Gilruth (director of MSC) to Fred Singer (director of the National Weather Satellite Center at the U.S. Weather Bureau) confirms the desirability of photographic work from handheld cameras at senior levels of NASA and other government organizations.[48] Despite Schirra's trouble taking photographs, Cooper reused the equipment on *Faith 7* for the same experiment with some additional filters for infrared photographs.[49]

Beyond weather photography of Earth and the mapping photography done of the Moon for the Weather Bureau and USGS respectively, the biggest non-NASA government "customer" for photography was the DoD. Capturing photographs for national security required that the processing and distribution of those photos did not unintentionally reveal more information than permissible by national security agencies. Historians James David and Dwayne Day have written extensively on national space security needs at the time as they relate to NASA. David explored the topic of astronaut photography and national security in a 2006 article in which he describes the reaction of those in the defense world to images taken of Israel and Egypt during Gemini missions, a particularly sensitive region at the time. Over the course of Project Gemini, eleven of fifteen experiments were carried out successfully with the "D" designation, meaning for the DoD, but only three included photography and were all completed on Gemini V. As part of the Gemini Summary Conference, USAF representative Wilbur Ballentine presented on the results of the experiments, and how they bettered the DoD's understanding of how human work in space might benefit his department.[50] Concern about the ability of astronauts to capture sensitive locations on Earth in high resolution continued during the Apollo program, mostly influencing the size of lenses used on Hasselblad cameras and the procedures for processing films. As most photography during Apollo focused on the lunar surface, however, it was not until Skylab that national security issues arose as a major issue for NASA. In fact, recent declassification information on Area 51 included verification of the DoD and CIA role in attempting to suppress astronaut photographs of the desert location taken during Skylab missions.[51]

The limited scholarly material on the subject of overlap between the intelligence community and NASA's human spaceflight program reveals how astronaut photography was closely scrutinized by the DoD and intelligence community for possible use.[52] The work of those agencies also influenced technology used during the space program, even into the Space Shuttle era. The issue of camera lenses and image resolution created problems for scientists wanting clear views of both the Earth and Moon. Concerns existed that the mapping and panoramic cameras proposed for the final three Apollo missions could reveal the capabilities of sensors and lenses used on National Reconnaissance Office (NRO) spy satellites, which were manufactured by the same companies.[53] As a result, Apollo cameras placed in the service module had somewhat lessened capabilities with a resolution of about three feet from seventy miles altitude. The film returned from the Apollo 15, 16, and 17 missions became the basis for the first high-resolution topographic maps of the Moon, and an invaluable resource for geographers and scientists involved in lunar research.

Mapping and Planning

Photographic work during Gemini and Apollo not only provided information for managers, engineers, and scientists, but also for mission planners. Images from completed missions could influence flight patterns or procedures on successive missions. That usually meant better, more efficient equipment and procedures or capturing different geographical features and landmarks along the spacecraft's trajectory. For Apollo in particular, imagery played a crucial role in informing scientists and engineers about the terrain of the Moon and possible landing sites. Such "bootstrap" photography became one of the most discussed topics in planning for missions.

As part of their postmission process, crew debriefings included questions about photographic work as part of the lengthy sessions done over days and weeks. Based on their flight experiences, astronauts often recommended modifying equipment, both large and small, to improve mission results. In the case of cameras and photography, suggested changes to the spacecraft windows were common. A notable change was required after Apollo 8 circled the Moon in late 1968. When three of the five spacecraft windows fogged up, making observations difficult, investigations revealed that a material used to seal the windows off-gassed in space. Over the next few missions, engineers experimented with different sealants to correct the problem.[54] With clear windows for handheld photography, mission planners used each mission as an opportunity to gather photographic data using the standard Hasselblad camera equipment and other specialized cameras as the Apollo program matured.

While NASA engineers utilized movie footage to develop or modify future practices, for the most part, capturing still images was a major part of mission plans during critical maneuvers. For example, in preparation for the first tests of the lunar module, one contingency plan included sending astronauts on a spacewalk to move from the CM to the LM if they were unable to dock the two vehicles properly. Plans also included adequate photographic coverage of such a process, including still images, to show how the spacewalking astronaut moved from one craft to the other.[55] To alleviate any fear that such photography might get in the way during the "Plan B" scenario, Deke Slayton, the memo's author, stated explicitly that camera use was on a noninterference basis. In other words, documentation of any kind, rightfully so, was secondary to astronaut safety and mission success. For example, the maneuvering done by Neil Armstrong on Apollo 11 to avoid numerous boulders and craters encouraged FCSD planning and training for Apollo 12 to include simulated landing photography. Oblique or approach-angle images improved the ability of astronauts Conrad and Bean to locate the correct crater where Surveyor III rested.[56] Other procedural changes included adjustments and notations for special photographs such as those of Earth and the Moon during the transit phases of the mission. As spacecraft window orientation during those flight phases often obscured views of the two, altered flight plans included procedures to reorient the vehicle in order to capture the appropriate frames.[57] Timing was critical for photography, so the desire to capture a full Earth image meant that only the Apollo 17 mission's outbound path to the Moon provided an opportunity for such a sight[58] (Figure 2.7).

One of the planners' highest postmission priorities during Apollo was to acquire and analyze photographs taken to adjust designated mission equipment, procedures, and maps for the next scheduled mission.[59] Layered with available lunar images from the robotic Lunar Orbiters and Surveyors, orbiting piloted vehicles became bases for surface geographical photography. Such imaging served to narrow down potential targets for human landings and provided photographic information for geologists at the USGS who created maps of the lunar surface. During the orbital portions of early Apollo missions, astronauts used bracketed cameras with timers attached to capture strips of images. For Apollo 14, a handheld topographic camera based on those already in use in reconnaissance satellites at the time provided additional images for developing accurate maps by lunar geologists[60] (Figure 2.8). On the final J-type missions, a complex equipment suite located in a special section of the service module provided data and images for research (Figure 2.9). The data retrieved from that system allowed the USGS to create the topographic images and maps of the lunar surface.[61] Planning for such photography was, in terms of

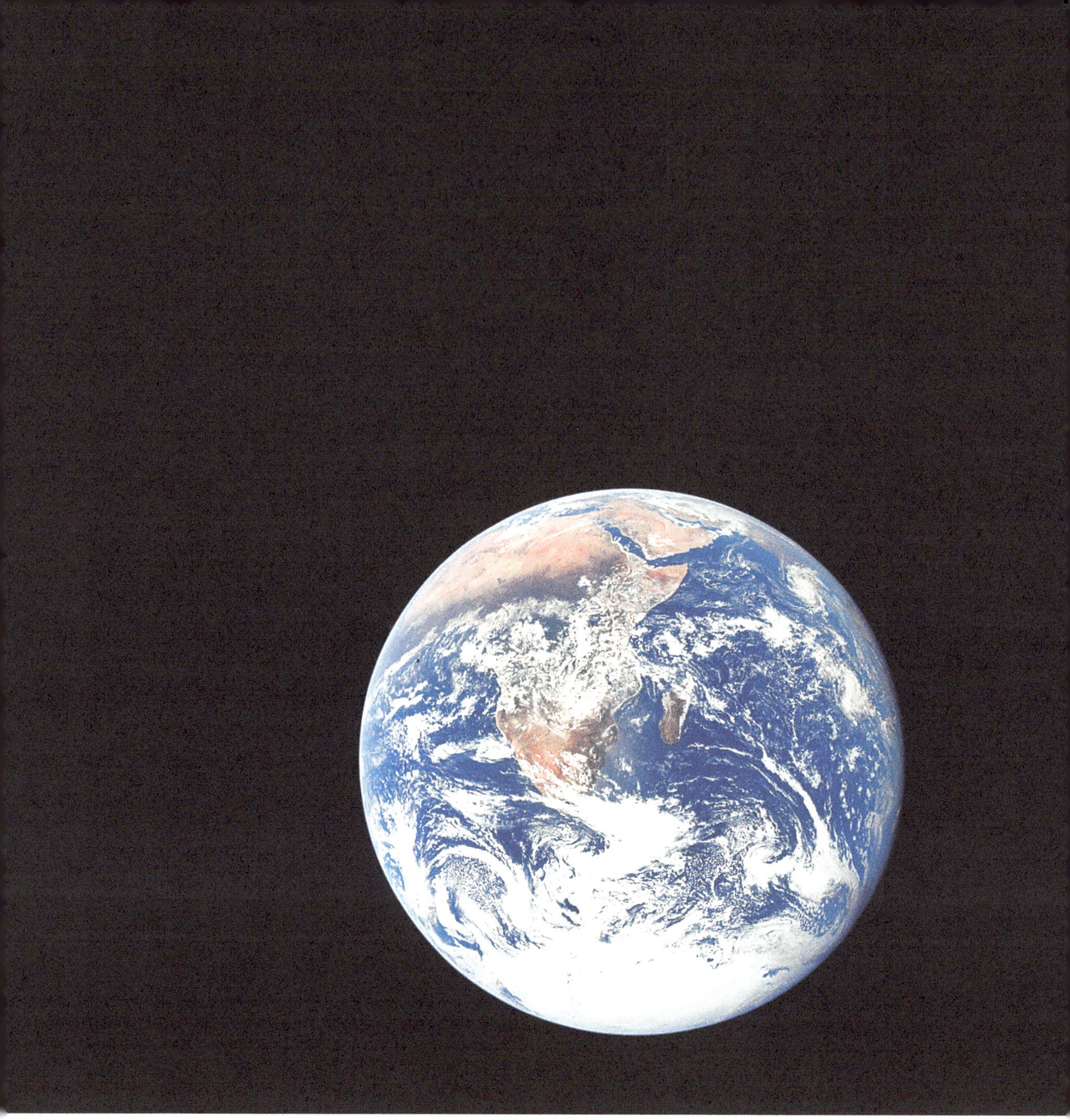

Figure 2.7. A whole Earth as seen during the Apollo 17 mission, likely photographed by Harrison Schmitt, AS17-148-22727 (NASA).

both the technology and the work, added seamlessly to the training and mission performance of the astronauts, with little creativity or response required from them to carry out requirements. According to Al Worden, Apollo 15 command module pilot, "These cameras were added to our flight plan to enhance our understanding of the lunar surface in a much more detailed way than ever before. . . . These cameras were mostly technical and I followed the instructions in the flight plan to operate them. Every operation of these cameras was prescribed before flight."[62]

Bootstrap photography, images made for the purpose of surveying later Apollo landing sites, often called for adjustments to flight plans.[63] For Apollo 12, a planned trajectory was altered in the months prior to flight based on the need to photograph

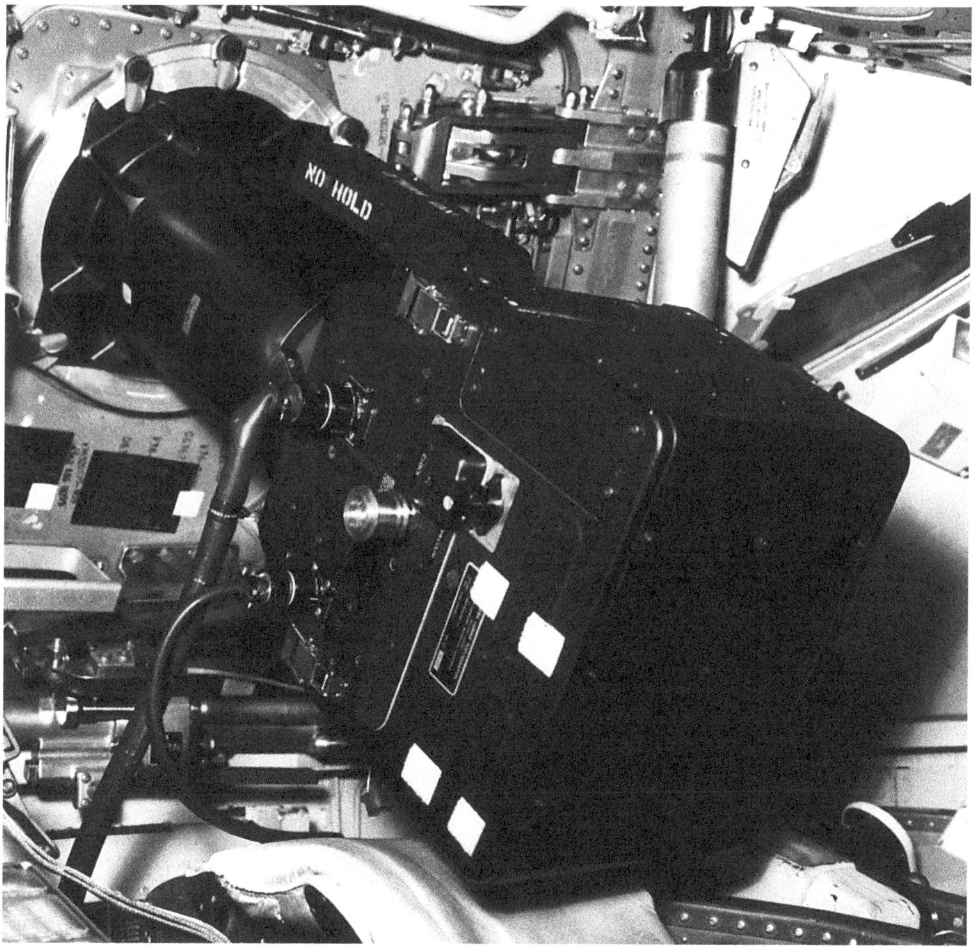

Figure 2.8. The Hycon lunar topographic camera was intended for lunar mapping during Apollo 13, AS13-282A-70HC-251 (NASA).

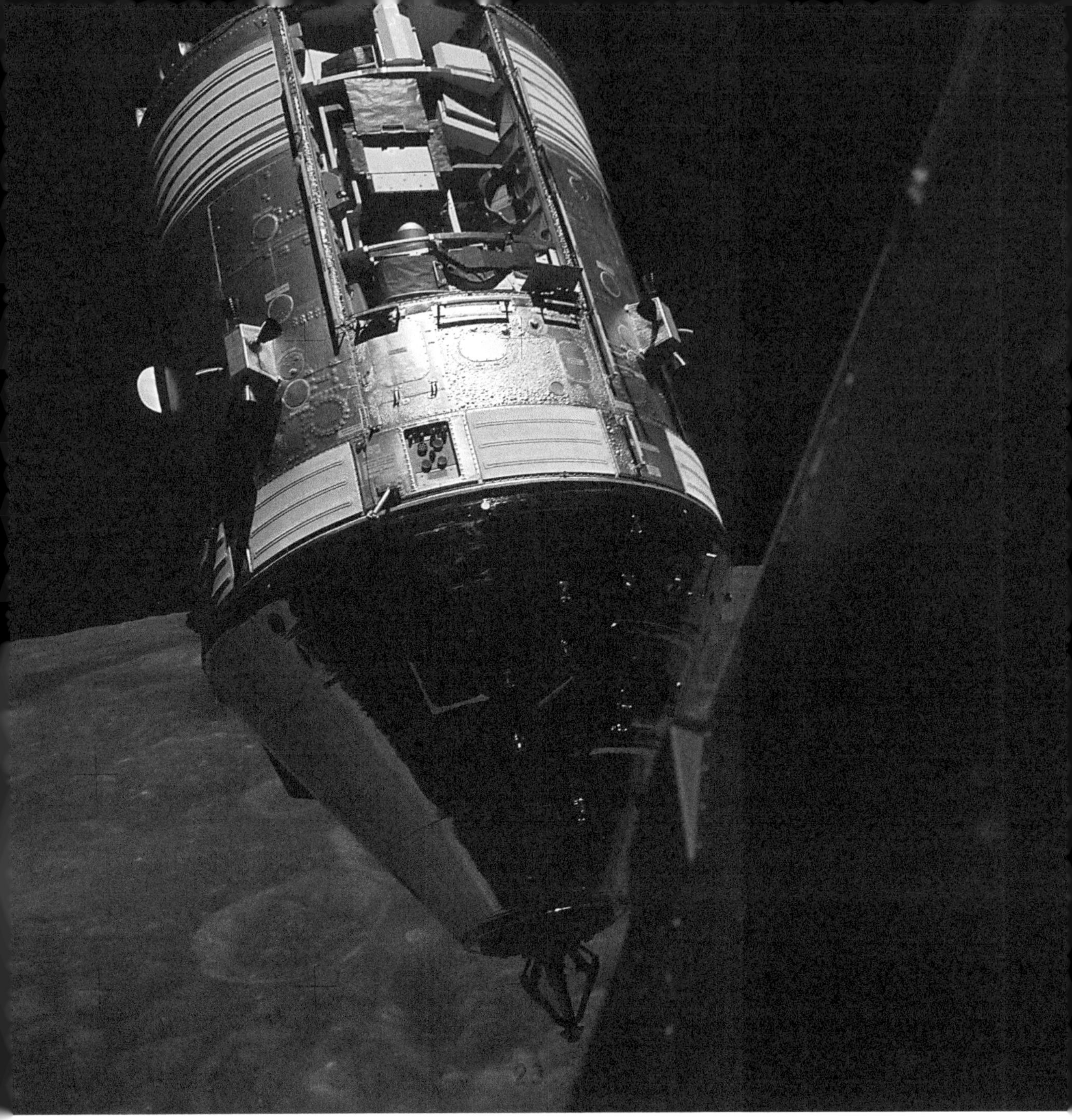

Figure 2.9. Open SIM Bay of Apollo 17 service module, December 14, 1972, AS17-145-22254 (NASA).

the projected landing sites for Apollo 13 (Fra Mauro) and Apollo 14 (Descartes), requiring an adjustment to printed onboard maps.[64] Such changes to flight maps and plans continued throughout the program, with significant input coming from scientists at Bellcomm, the contract managers for mapping photography and home to noted geologist Farouk El-Baz. As a geology instructor to the astronauts, particularly the command module pilots on the last three missions, El-Baz trained them to verbally describe and pinpoint with cameras the necessary landmarks on the Moon.[65] However, such work was limited and received no more time on the flight plan than those managing crew time felt was appropriate. Pushback on the workload came in the form of rejections for additional photographs from FCOD when a flight crew's schedule appeared too full and passed a threshold for daily activity.[66] With the photographic data analyzed by scientists and managers, site selection committees made the final decision regarding landing locations, but available documentation shows the high value placed on the photographic work of astronauts for planning and mapping purposes.

Public Affairs and the Media

The conduit for outwardly communicating NASA's work was the Public Affairs Office (PAO) at Headquarters in Washington, D.C., and at each of the centers around the country. This office not only represented NASA to the public, but also brought agency engineers some sense of what visual materials would interest the public and best convey the agency's message about human spaceflight. While program scientists and engineers involved could directly influence photography from within the agency, or through principal investigators, the media and the public had only the PAO for visual information about the dangerous and costly missions. The PAO, under tremendous pressure to explain every possible scenario, considered all points of the mission as a possible instance for conveying the NASA narrative.

From the outset, NASA managers considered the relationship between the public visual representations of spaceflight. Those on the inside of the process, the people responsible for procuring films, cameras, and other equipment, spoke at length about the value of photography as part of the public service dimension of exploration. John Brinkmann voiced his opinion about photographic work in a memo to the chief of the Systems Engineering Division upon the occasion of the cancellation of a lunar mapping camera project. Responding to a mid-1968 change in his department's role in developing films, Brinkmann pinpointed the need for a variety of still image types: "Film is light in weight per unit (about 1 gram) and has a great scientific impact per unit. There is a possibility that it could be the only item of importance evolving from the mission, and it is the only tangible contact that the general public has to the program."[67] For a professional photographer and manager to show an awareness and concern for public need reveals an appreciation for the importance of images perhaps not even fully understood by public affairs officials.

In the early stages of planning during the Apollo lunar landings, the MSC PAO wrote a lengthy list of required photographic coverage elements to Joseph Shea, then head of the Apollo Spacecraft Program Office.[68] The four-page memo included general criteria for motion picture shooting, to occur in parallel to 70mm photography. The all-encompassing list of documentary photography, which Deke Slayton's Flight Crew Operations Division deemed unreasonable and extreme, included nearly 100 different elements that the PAO intended to make available to anyone that might request images. Hardly a moment of the mission went unconsidered in creating the list, including storing empty food pouches, climbing down the LM ladder, and opening of the main parachutes for the water landing. While there were three crewmembers to capture some of these requested moments, one might imagine that the circumstances of spaceflight might make it difficult to pull out a heavy camera to take a snapshot.

Advocates for historical and documentary images did exist outside the PAO, suggesting the importance of connecting the public to the experiences of Apollo astronauts. In a particularly stinging memorandum to the chief of the Mission Operations Branch, noted engineer and the father of the blunt-ended capsule design—Max Faget—criticized an early draft of the Apollo 11 lunar landing photographic plan for its near complete disregard for "the visual appreciation of the astronaut's activity on the moon during this singularly historical event."[69] Faget continued with his unfavorable appraisal of the plan by pointing out a lack of intent to photograph both crewmembers with the Hasselblad camera, using the terms "stingy" and "low quality" in reference to the overall plan content. As the chief of the engineering and design division, Faget was also keenly aware of the need to satisfy the short- and long-term impressions possible with still photographs. In perhaps a more telling memo from later in the program, scientists in NASA's planetary and Earth resources division made a suggestion one might assume to be an obvious one to have made at the start of the program. Paul Gast, the division chief, wrote to the JSC PAO to say that the two units should work together to ensure that film frames were used efficiently for both their purposes, then processed in a manner to support their needs with the public and those of the geologists studying samples taken.[70] Another motivation for this recommendation was the possible input of the scientists into the release of photographs, with Gast stating their value in "making available to the news media photographs that are more meaningful than randomly selected individual shots."[71] In the cases of engineers and scientists, seemingly common sense approaches to planning photography in conjunction with the PAO fell at rather late stages in the program, perhaps revealing a poorly thought-out plan for visual communication with the public.

Corporations

Corporations could benefit from information contained in astronaut still photographs, a possibility they themselves recognized by end of the Mercury missions. Within months of Gordon Cooper's *Faith 7* mission in May 1963, at least two companies, Mobil Oil and ITEK, contacted NASA regarding ways to integrate input from astronaut photographic experiences into their work. In the case of Mobil Oil, researchers at one of their laboratories contacted NASA to see how they could analyze Mercury photographs for what was presumably any determinable information in the photographs on oil deposits.[72] While photogrammetric analysis was already underway at the Goddard Space Flight Center on the data, sharing the results with a corporation for their potential gain posed no concerns for NASA managers as the photos and any data gleaned from them was a matter of public record and scientific interest.

In a very different way, a photographic technology company wished to use information from NASA to inform the way it manufactured and tested its products. From the experiences of Mercury astronauts, ITEK, a large format camera maker from Palo Alto, California, hoped to learn how it could improve its own company-funded research into the production of a lunar reconnaissance camera.[73] While the camera they proposed evolved into the remotely operated camera flown on the last three Apollo missions, they felt at this early stage of development that the astronaut point of view captured via handheld cameras could inform their investigation. ITEK sought information well in advance of its eventual contract to create the mapping camera, but its interest in the human need for such images built upon its existing work on remotely operated spy satellites for the U.S. intelligence community.

The commercial implications of astronaut photography may appear hazy at first glance, but companies who dealt with Earth resources issues such as oil, gas, and other mining operations benefited from the earliest color images taken by people from space. Those operations turned to the cheaper and safer method of satellite imaging in the later 1960s and 1970s with the rise of programs like the Nimbus, LANDSAT, and GOES satellite systems.[74] Mercury astronaut photography, however, contributed some of the first clear color Earth images to scientific work done to steer corporate operations. By the end of the Apollo program, LANDSAT's emergence as a potentially commercial image program ended the specific need for astronaut photography for mapping. In the end, the attempt to commercialize the resource failed and it remained freely available to anyone interested in Earth-directed photography.

The most obvious commercial use of astronaut photography during this period was the retail market for ephemera related to human spaceflight. John Brinkmann, whose office was ultimately responsible for processing visual materials from flight, identified a forthcoming demand for images following the lunar flights in particular. He expressed deep concerns over the process for distributing films to commercial entities in a memorandum sent immediately before the Apollo 11 mission.[75] While there was no argument against the legality of companies producing films and other products for profit using NASA visuals, Brinkmann expressed serious concern over the workload such requests would create and the lack of a defined procedure for handling requests. Above all, he sought to ensure no other uses would make it appear as though those companies had exclusive rights to use or distribute the images—they were all public domain. NASA had—and still has—no ability to stop companies from creating retail products based on images taken by astronauts, but they could control the official agency descriptions and the perception as originator of, and authority on, astronaut photography.

The Public

As a public, government-funded entity, NASA answered to the largest audience for astronaut still images—the American taxpayers. What those images meant to the public requires visual and cultural analysis, but NASA presumed to know the needs of the public audience during its early human spaceflight programs. I would contend that the most direct means for the public to understand the physical reality of spaceflight, and to continue supporting the project, was to represent humans in the context of space via still images. Statistics show that millions around the United States and the globe watched live and recorded coverage of NASA missions on television, but still images, for their ease of reproduction and distribution beyond the reach of television, became the primary means by which most people viewed NASA's activities. This required a procedure for those images to go from film magazine to publicly available resources like newspapers and magazines.

One public audience, far larger than the immediate beneficiaries of any of NASA's research and photography, was the international community, including those in positions of authority and anyone with access to printed news media. International requests for photography were recognized as opportunities for goodwill from the leadership level down through to those who fulfilled the requests for the images. Lyndon Johnson even used the *Earthrise* image as a reminder to world leaders what good the U.S. government brought to the world.[76] As a matter of daily work concerns, the PTL bore responsibility for fulfilling requests from inside and outside the United States for reproductions of astronaut still and moving images. Following the Apollo 8 mission, Richard Underwood expressed concern about a lack of clear procedures for working with NASA's Office of International Affairs (OIA) and PAO to satisfy the needs of international requests for images. In resolving such deficiencies and streamlining the connection between the PTL and OIA, he identified the benefit to NASA as one in which "photographic programs have a great potential of establishing good will with our international neighbors."[77] As a means of communicating a positive message about human space exploration by NASA, and thus the U.S. government, photographs by astronauts—our surrogates in space—became critical tools for encouraging positive foreign relations.[78]

The public had few mechanisms for influencing NASA's process for prioritizing photography, and without knowing what would satisfy their curiosity about spaceflight, most naturally identified with familiar representations. Many still images featured some element of the spacecraft or other spaceflight technologies, but that did not mean viewers outside the engineering or scientific community would understand the photograph's content. As an audience for astronaut photography, the public needed guidance from NASA and the astronauts. That assistance came through the voice of public affairs officials and news media interviews with astronauts, who played the role of interpreter to a public desperate to comprehend image content. Understanding NASA's process of gradually zeroing in on images easy for viewers to understand required non-astronauts interceding in the after-mission life of the images, shaping the legacy of astronauts and the pictures they brought home.

PHOTOGRAPHIC OPERATIONS

Audiences for astronaut photography had varying degrees of success receiving images because of their direct or indirect expressions of need. The process by which those images made it from film roll to viewer shaped their consumption. NASA developed a procedure for moving all flight films along this trajectory, primarily though the work of those in the PTL. Unfortunately, any records generated by that office appear lost or destroyed, but certain memoranda retained by other NASA offices regarding the PTL's process and relationship to the PAO, most clearly documented during the Apollo program, illuminate the ways in which still photographs became a usable commodity for intended and unintended audiences.[79]

Upon return of spacecraft to Earth, recovery operations brought the vehicle and its occupants from their ocean landing to an aircraft carrier. As the spacecraft and its occupants were unloaded after nonlunar landing flights, a representative of the PTL picked up film magazines used during flight and returned them to the lab for processing within thirty-six hours of a mission's end.[80] In an outline of procedures for film retrieval for the Apollo 8 mission, the chief of the Test Division wrote that "Any onboard photography containing earth-looking views shall be prepared for distribution at once, but shall not be released until so authorized by the Director, MSC."[81] This indicates a pause in process of getting still photography released that allowed for a national security review.[82] Written for the Apollo 8 mission, the enclosed outline with the memo details the distribution of still images throughout NASA and the timeframe in which that was to occur. Listed first for each type of image to distribute is David Goldenbaum, a member of the Apollo Spacecraft Office's Test Division. His position at the top of each category was explained with a note that his copies were for the MSC Mission Evaluation Team.[83] Implied within this prioritization (which neglects to include the initial national security review period) is the need to analyze visually every aspect of Apollo missions, both to supply information to those who would speak about the mission publicly and to inform future missions.

Film magazines from the first three successful lunar landing missions were processed somewhat differently. Once the spacecraft, its contents, and the astronauts arrived in Houston, everything went into the Lunar Receiving Laboratory so that any potential lunar debris was contained in an environmentally controlled space.[84] Postmission processing for Apollo missions 11, 12, and 14 included concerns about the possible transmission of "moon germs." Quarantine was established in the LRL, which covered staff involved in processing the spacecraft and its contents, those who cared for the astronauts, and the astronauts themselves during the week prior to landing and up to twenty-one days postflight. One staff member involved in processing films for Apollo 11 (and other missions), PTL technician Terry Slezak, offered valuable insight into the methods used to decontaminate film magazines and transfer them outside quarantine for distribution in his official NASA oral history. In addition to photographing astronauts inside quarantine, Slezak also decontaminated film magazines with ethylene oxide gas without exposing the film.[85]After completing that process, Slezak explained that armed guards arrived with a PTL courier to rush the magazines to the lab for processing and distribution. The exact procedure was simulated more than a dozen times just like any other NASA process during the year leading up to flight films returning.[86] Slezak gained notoriety for having touched lunar dust by accident during his processing of Apollo 11 film magazines when he picked up one that Armstrong accidentally dropped on the Moon (Figure 2.10).

Processing flight films at the PTL underwent its first test after the Apollo 7 mission. According to lab chief Brinkmann, staff went on twelve-hour shifts for the first six days to process moving and still films for those getting the first set of duplicates from the original film rolls. Processing film on this round-the-clock schedule apparently set them back on preparing for Apollo 8 only two months later, since the lab was responsible for loading magazines prior to flight as well.[87] With a small staff, the PTL contracted out some of this work, and the experience of Apollo 7 prompted Brinkmann to write his superiors to alert them to a potentially disastrous situation with increased films returning from future missions. As seen in Table 2.1, still frame returns went up significantly with each successive mission, which gave the lab increasing work and a requirement for new equipment and more technicians.

The route taken by scientific films, those taken with special film or using filters that made them useful only to scientists, however, is far murkier for historians to discern. Numerous scientific proposals undoubtedly arrived at NASA for

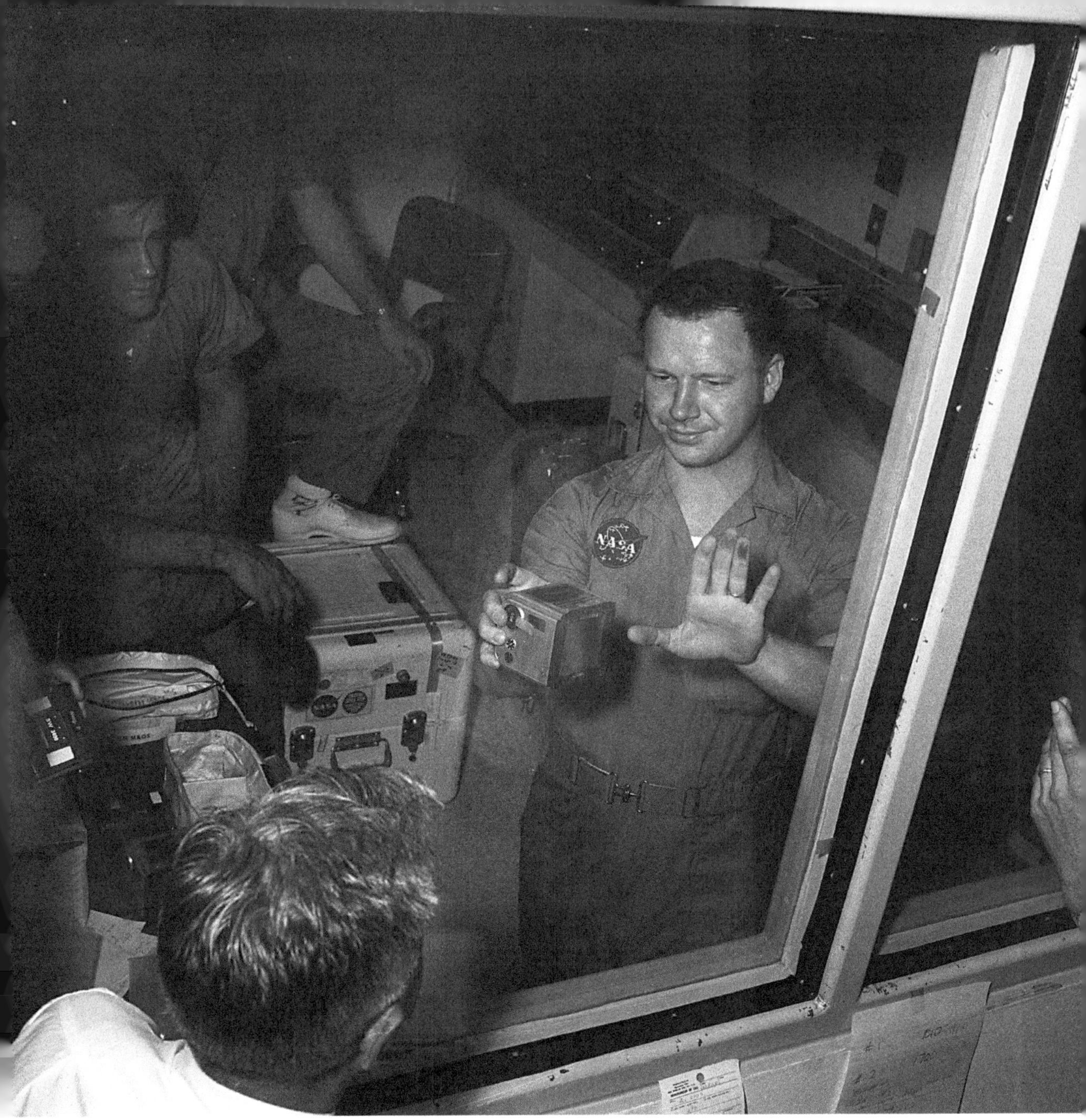

Figure 2.10. Terry Slezak in the LRL with an Apollo 11 film magazine and lunar dust on his fingers, S69-40054 (NASA).

experiments. Flights to low Earth orbit and beyond remove atmospheric interference from observations made through cameras and telescopes, for example, making spacecraft a better platform for astronomical and other scientific studies (though not as good as a robotic vehicle). As the data acquired for these experiments was visual, astronauts used special film magazines, designated for specific experiments, to capture the appropriate shots. Special processing for these films frequently occurred at contracted labs with specific capabilities for ultraviolet film development, for example. Numerous resources exist on the experiments by way of mission reports, post-program scientific conferences, and results reports required of the principal investigators.[88] The visual results, however, are often either unavailable in a digitized format from NASA, no longer in NASA hands, or were deleted from NASA archives. In an attempt to acquire images found in the Gemini V mission report, taken during an experiment to photograph the zodiacal light phenomena, I discovered that those images were deleted from the JSC image library. How or why this occurred is unclear, but it may be possible to track the images down via the records of each principal investigator. As scientific and astronomical images are not my focus, locating them and researching the reasons they no longer exist in an official NASA repository remains a topic for future research.

The long-term accessibility of Hasselblad orbital photography for lunar research purposes is worth discussion. Because scientists remained the leading audience for astronaut still imaging, entire sets of reproductions just a generation or two removed from the flight films found homes with those most in need of images for their research. Through the end of the Apollo program, NASA remained the only repository for images needed by lunar scientists through its relationship with Bellcomm. The company maintained a library of images and technical information for use by their staff and NASA. When Bellcomm ended operations with Apollo's end, the entirety of the image and reference collections as well as some staff transferred to the Smithsonian National Air and Space Museum, becoming the nucleus of what is now the Center for Earth and Planetary Studies (CEPS). Farouk El-Baz, Bellcomm's trainer to the astronauts in lunar mapping and orbital photography, became the first CEPS director and helped steer CEPS toward involvement in astronaut photography of Earth during the Apollo Soyuz Test Project in 1975. A few years later, scientists and other researchers acquired the ability to access the same high-quality photographic reproductions of astronaut still photography through a larger set of facilities established as part of NASA's Regional Planetary Image Facility network, the first of which was CEPS. While the digitization of images renders onsite research at these facilities somewhat antiquated today, the principle of wide distribution and accessibility to the community of interested scientists originated at NASA at a time when the photographs required hard copy libraries at nationally and internationally recognized lunar and planetary research facilities.

The final step in taking film magazines from flight to the eyes of audiences other than scientists, at least in the case of photographs deemed relevant for immediate public release, was the addition of captions. These content messages attempted to frame conceptions of the images, though little of the text came to linger in the memories of viewers. My research into historical PAO files did not uncover documentation of how this occurred, but the oral histories of Terry Slezak and Richard Underwood explain the involvement of other staff in generating the textual information. As a means of communicating the details and context of the image, the caption provided by NASA was the most common way the media and public viewers received additional information about the image. For Gemini missions in particular, which all remained in Earth orbit, Underwood developed his own process for comparing mission transcripts, magazine numbers, and maps to determine exactly when in the mission timeline a certain image was taken in order to identify the content and write up descriptions, for either scientists or public affairs officers.[89] Underwood did not explicitly state that the same process, perhaps with assistance from lunar scientists, continued through the Apollo program. Terry Slezak commented in his oral history that "we interfaced with the astronauts a lot. They would pick out certain pictures from their missions that they thought were the most salient features of the mission."[90] With explanations from the astronauts, Underwood's research, and input from geophysical or lunar scientists, descriptions of image content were developed to inform those lacking knowledge of spacecraft, Earth, space, and lunar features.

Who then decided which images of the hundreds or thousands available would become available to the average viewer? Slezak's comment indicates that the astronauts did play a role in guiding PAO officials in which images best highlighted the significant moments of their missions. Richard Underwood said, on the other hand, that "Public affairs would get the ones they liked, and I'd give them a nice big write-up on those pictures so they could release them," which positions him as a crucial voice in conveying information about image contents. Almost certainly, reality rested somewhere in between both scenarios, bringing together the knowledge, ideas, and information held by the astronauts, scientists, public affairs staff, and that of photography specialists. Non-astronaut NASA staff bear some degree of responsibility for the agency's visual character as informed by evidence collected by astronauts. Those images are cited even today as proof for and against NASA rhetoric regarding human spaceflight.

Expectations Fulfilled

While the types of images astronauts captured were somewhat predictable, rarely if at all did those at NASA consider openly how publicly released images might shape perceptions of this heroic period of exploration. The creation of a public understanding of NASA's mission in human spaceflight required the involvement of administrators, public affairs officials, and the astronauts themselves, using images that also fit an existing visual rhetoric of triumph, technological achievement, struggle, and humans as part of the landscape. Moving and still images gave audiences the evidence needed to complete processing of the personal testimony of their surrogates in space exploration. Each audience then had the opportunity to consume astronaut photography for what information they needed.

They were well trained in photographic plans and had some of the best quality camera equipment available for multiple uses. Mission plans acted as scripts so astronauts could easily infer the desires of interested audiences, especially those at NASA and in the scientific community whose instructions were very specific. Still, astronauts acted as photographers and used their own sense of composition and subject when veering off schedule for composed shots that conformed to the existing visual tropes and remain the most common in the visual lexicon of the space age. Astronauts, using the guidance of the written plans and using their own instincts, knew what made for a good photograph. Their results showed that crew-selected images often fulfilled the simple public needs and expectations. The thousands of still images captured for engineers, scientists, other government agencies, and corporations satisfied those requirements, but as evidence, they were not a significant means by which the public constructed a collective memory of the experience of human spaceflight.

As part of our collective understanding of the experiences of exploration, certain images, perhaps best described as icons of the space program, endure as symbols that undeniably recall a sense of accomplishment, success, and pride. Classic landscape views, horizon or "frontier" studies, astronauts demonstrating pride in their achievements, technologies of travel set in the alien environments of space, and the enormity of terrain features of the Moon were what situated astronaut photographs comfortably within a visual history of exploration. Along with perhaps a half dozen other images cited here, most featuring astronauts in the setting of space, make up the small set of familiar images used consistently to represent the heroic era of U.S. human spaceflight. Knowingly or not, NASA asked astronauts to provide images that fit a series of visual tropes of exploration that existed prior to humans venturing off the Earth's surface in the twentieth century. So, while the images captured ultimately fulfilled the rhetorical needs of NASA, they were far from representative of the entire visual record of human spaceflight.

CHAPTER 3

Images of Exploration

> *"I know of no reason why photography should not find favor with the pioneer whose object is to map out a new route and to picture to the scientific world at home, in a trustworthy manner, what he himself has observed during his travels."*[1]
>
> *—John Thomson, Photography Instructor, Royal Geographical Society, 1891*

Walking in space or on the Moon inevitably generated some very unique stories. Photographs were the primary means, beyond hearing directly from the astronauts themselves, of learning about what they experienced. Only the truly dedicated space enthusiasts read mission or debriefing transcripts, consumed television or radio interviews, or read enough articles to get a comprehensive understanding of NASA's human spaceflight program. Astronaut stories come alive and seem understandable when illuminated by the images captured, supported by captions and personal descriptions. These photographs seem familiar in part because of past exploration, which set up a visual precedent for how we see exploration generally. Photographic documentation of places such as the American West and Antarctica, locations explored in the late nineteenth and early twentieth centuries, serve as visual analogues for human voyages off Earth as well as training locations for their work while still on the Earth. How those images were created, the visual rhetoric they established for exploration, and how the public interpreted the embedded messages of exploration photography offer a model for analyzing NASA astronaut photography from the 1960s. While not an overt part of the planning for astronaut work, the questions needing answering and themes to be addressed by astronaut images bear a striking resemblance to historical analogues in extreme environments on Earth.

In the case of lunar exploration, few images came to represent the experience and accomplishment of human spaceflight like *Moonman*, featuring Buzz Aldrin on the Moon as captured by Neil Armstrong (Figure 3.1). Images such as this or Apollo 17's view of the whole Earth from 500,000 miles away were often duplicated and reinterpreted by artists and in popular culture and gained nicknames related to long-standing references to the content of the photograph. In this case, references to astronauts as "moon men" appeared in countless newspaper articles at the time. The Aldrin image ultimately inspired MTV's music award statuette and forever linked the name and the image.[2] As one can do with an image of the flight of the Wright brothers near Kitty Hawk, North Carolina, a close examination reveals interesting details that enrich our understanding of a sublime moment in history.[3] Through the history of exploration, in fact, we can read images for a visual rhetoric, embedded messages that provide the basis for justifying and defining why we explore, that later inform and inspire exploration and cultural products. It is possible to examine that visual continuity by comparing, side-by-side, the images of exploration from three distinct yet similarly photographed exploration projects, and then scrutinize our shared cultural memory of the images.

Even without documentation or interviews with Aldrin to rely on, we could infer from the image what he was doing at that moment—an image reprinted, reused, and reinterpreted many times over since 1969. Reading the image, we see Moon rocks, soil disturbed by boots, and a portion of the landing pad and the landing probe bent back upon impact. Then there is the astronaut, Buzz Aldrin, his golden visor reflecting the photographer and the Earth, dirt-covered knees, and a pop of color from the stitched-on American flag on his shoulder. However, there is movement implied by the photo: a strap dangling from the front of his suit caught mid-swing, a crooked arm, and his left toe dug into the lunar regolith. Interviews following the mission reveal minimal details of that moment: Armstrong thought to get this unplanned portrait of Aldrin from a distance. Though not documented clearly in mission transcripts, Aldrin was moving to look at his wrist checklist and still adjusting his stance based on the weight of his spacesuit backpack. This deeper look into a single image enhances an inspiring view of human activity on the Moon. Seeing a person in that place and being able

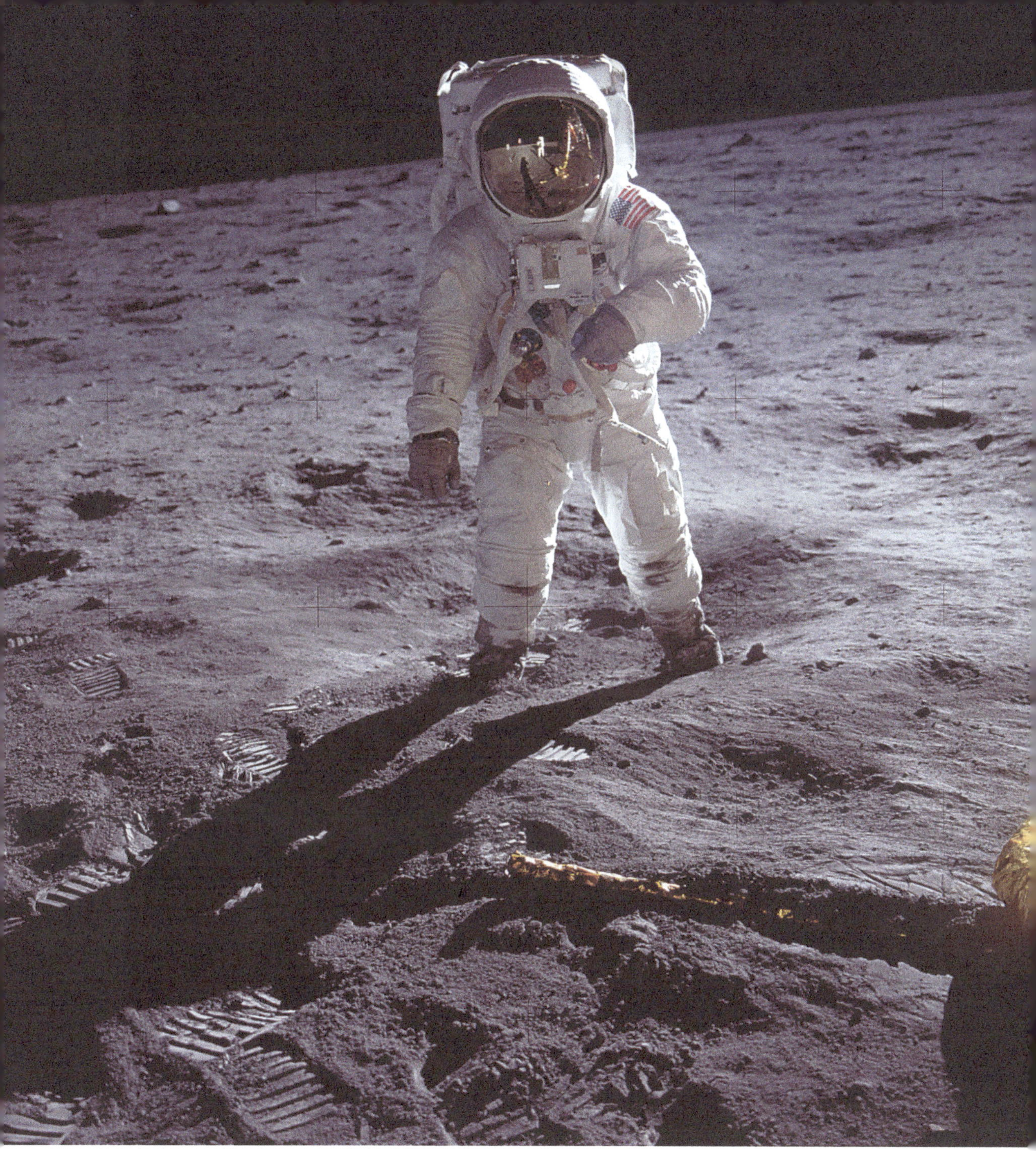

Figure 3.1. The iconic photograph of Buzz Aldrin, over many years, came to be known by many as the *Moonman* image, AS-40-5903 (NASA).

to imagine the experience is a particularly powerful tool in creating understanding among non-voyagers. NASA did that with its photographs in the tradition of exploration of hostile environments on Earth. The location may change, but the messages from over one hundred years of exploration photography are often surprisingly similar.

NASA's early astronaut photography presents one story of exploration. It was by far the most visible project of its kind, permeating American culture though easy transmission of information by television, radio, and print media. It featured the most common styles of photography: landscapes, documentation, and portraiture. Humans appear in these images largely as points of reference for the viewer, not as subjects of study. Apollo 12 and Skylab 3 astronaut Alan Bean, an artist in his own right, lamented the lack of images containing astronauts in his book about astronaut photography: "We didn't emphasize the human aspect enough, and I tell you it was a mistake."[4] I assert that the absence of personal, familiar, or identifiable objects in images for the public to connect to contributed in some ways to the failure to build long-term support for NASA activities. Seeing our surrogates participate in such spectacles made their work real and exciting, worthy of support as they instilled pride and a sense of accomplishment during dark days. The majority of exploration images are landscapes or spacescapes. Nearly all of those that became icons of the program and the heroic era of spaceflight feature a human element, though sometimes in the abstract (e.g. boot prints, shadows, human-made technology). The term *landscape* implies the presence of humankind, and the act of photographing it creates an idea of land encountered by people.[5]

Expedition members who captured images play the role of surrogate: their perspectives, views, and eyes act as substitutes for our own, bringing us sights that delight, surprise, and inform. In place of audiences participating in the journey, photographers provided a visual sense of places nearly impossible to visit otherwise. As case studies, exploration of the American West and Antarctica provide equivalent scenarios in terms of goals and environment consistent with space exploration, so I will first look at those cases in some detail to establish common elements of production and distribution, then circle back to how those are expressed in images and were understood by audiences. Exploration projects such as western European empire-building in Africa, oceanography, and mountaineering share some common themes with space, but the West and Antarctica share the most in common with astronaut photography in terms of visual rhetoric, goals, and public memory.

Whether intentionally or by happenstance, images take on meanings other those intended or expected by expedition sponsors, photographers, or audiences. By reviewing widely seen images from terrestrial exploration projects, their content and composition indicate a consistent set of themes that create a visual rhetoric, not only for each project, but also throughout exploration of Earth and space. Illustrating crucial messages, photographs from the American West and British Antarctic expeditions contain the seeds of visual rhetoric seen in NASA's human spaceflight program. These earlier projects provide exemplars of visuals captured to promote them: placing unknown locations in a human context, representing the means of transportation to these remote locations, grand and dramatic views, natural challenges to exploration, and scientific goals. Distribution and public consumption over time created a level of recognition of visual themes and expectation of their repetition by later explorers with cameras.

IMAGES FROM EXPLORATION

Some rhetoric, conveyed through words and images, was more effective than others in corralling wide support for exploration. When examining visual and textual languages used in each period, similar themes emerge and connect these projects across time, giving a sense of continuity to human exploration through dramatically different historical moments. Common goals for expeditions included scientific discovery, locating natural resources, increasing support for territorial expansion, and opening new areas to commercial and economic exploitation.

These exploration periods have common characteristics that require a comparative analysis using categories from visual culture studies: production, processing and distribution, and audience reaction. First, producers conceived of and promoted the images, selected still camera technology based on certain criteria, and had the surrogate photographers trained for producing images. Second, when images returned from expeditions, they required handling, processing, and distribution. It is vital to examine those elements in terms of camera use in the context of the American West and Antarctic expeditions. Narratives of image creation relied on rhetoric developed for photography's initial purpose to direct the post-mission process of distribution. The last stage involves the interaction between audiences and the images. No matter the goals set by photographers or expedition leaders, audiences responded to the images in sometimes-unpredictable ways. Analyzing audience reception, the next stage, becomes the means of discovering photographers' ability to make a lasting impression on public memory.

Producing useful images—documents of exploration—requires a steady and trained hand able to wield the image-making technologies selected. A major expense of any expedition is the payment of staff trained in tasks crucial to fulfilling the rhetorical goals of the project or carrying it out. The degree of professional training for photographers varied from one expedition team to the next, so any formal training or professional experience of expedition photographers influenced their role in the narrative of the expedition. As a burgeoning imaging technology and potential art form in the late nineteenth and early twentieth centuries, those claiming professional experience as photographers more often learned the craft through practicing in commercial settings, studying under studio owners or teaching themselves through practice and published instruction guides.[6] Rare was the instance when a formally schooled artist of the nineteenth century took up the camera with the intention of creating artistic images. It took many more decades for the camera to rise in the esteem of the art community. Photographers of early exploration—and astronauts of the mid-twentieth century—learned their craft through participation and practice. In contrast, though, astronaut training and practice was largely a sideline that some astronauts took more seriously than others. During their missions, they did not have more than a brief moment to consider artistic elements like composition, lighting, or indeed anything beyond operating the camera.

What separates the efforts of professional expedition photographers and the astronauts is the *intent* of the photographs they produced. For a professional photographer, the process of creating a photograph ranges from selecting the technology to affix the image to a plate or film to the display of the print. This is the creative process of an artist: producing an image and presenting it to viewers in an intentional way. Astronauts had no such creative intent or control. Their role in the photographic process was to collect images. They simply needed to understand how to manipulate the equipment, follow written guidelines on camera settings, and return film magazines safely to Earth. They created a collection of images, sometimes offering thoughts on interesting images or caption information, but that is where their part in processing began and ended. NASA staff, including photography technicians, managers, and public affairs officials, brought the process to its completion by developing films, handing them off to either scientists or another intermediary for distribution. We might consider astronauts as akin to census data collectors or pollsters. Their work required moderate instruction to perform the photographic task, removing them a few steps from consideration as professional photographers. This should not serve, however, to diminish the importance of their role, but to place their efforts along a spectrum of experience and motivation.

While bearing some resemblance to expedition photographers of the heroic age of terrestrial exploration, astronauts bore few of the same philosophical or artistic burdens of those men. The clearest connection between the stories of training for photographers during exploration of the American West and Antarctica is the lack of formalized art education. As members of the scientific teams charged with creating visual representations, survey photographers created narratives with framed images that needed to represent the most important aspects of the expedition. Their constituents—the expedition leaders, financial supporters, and general audience—needed satisfactory products and confirmation of the established rhetorical goals of the expeditions. While astronauts were dedicated to performing photographic tasks to the best of their abilities and accomplishing virtually the same visual goals as previous expedition photographers, they completed those without the luxuries of time or space. Environmental challenges faced by explorers are the most significant difference between the experience of photographic exploration of Earth and space with cameras.

No expedition or project moves very far without a mission statement or list of goals the participants wish to accomplish. While the people and language changed over time, the intent of the rhetoric remained much the same. Steeped heavily in terminology associated with competition, commercial opportunity, and scientific discovery, expedition leaders needed images to convince audiences of the utility of such efforts in hopes of garnering additional funding for more work. Whether from the project leader, the lead scientist, or the entire team, someone established the basic rhetoric that defined the work, its aims, and the benefit of a project to an audience. By definition, *rhetoric* is the use of words to please, influence, and persuade. In the case of exploration, charismatic and politically motivated individuals and groups developed rhetorical devices to encourage potential donors, excite audiences, and proclaim some measure of success despite any failures. Those who created the rhetoric, leaders of the project in one way or the other, used it to define what followed. From the justification of the Lewis and Clark expedition as an effort to locate a practical transcontinental route to the challenge presented by President Kennedy for a new generation of explorers, rhetoric played a significant though symbolic role in exploration. Wide proclamation of this rhetoric via scholarly publications, letters to potential funders, lectures, and the media—and the media served exploration leaders as tools to sell their ideas. Using easily collected images of explored lands bore out the fruits of those promises for the audiences more quickly, creating a visual rhetoric or evidence of successes, failures, and the struggle undertaken all around.

The conditions under which photography became part of NASA's mission to send humans to explore space are reminiscent of, and informed by, those from exploration of remote locations on Earth. Going back to early exploration of the New World, Africa, and the Far East by western Europeans, imaginative artwork accompanied the narratives of expedition parties and their harrowing stories of bad weather, treacherous seas, and confrontational native peoples. If their ships returned at all, with or without a bounty of goods acquired from the new lands, both textual and visual reporting played a role in encouraging governments and companies to invest in further discoveries. Mapmakers, scientists, and painters created visual representations for audiences, while heroic and charismatic explorers planned new expeditions bound up in promises of more support from financial backers.

In the exploration of the American West and Antarctica, the relatively unknown nature of those lands meant explorers developed a rhetorical means of explaining their goals. On the surface, these expeditions appeared scientific, but at their core they were efforts to strengthen national pride through competition and access to resources. Trekking across North America, or the entire Antarctic continent constitutes a large part of the activity undertaken in what many refer to as the Heroic Age of Exploration.[7] This period, marked by penetration into the interior of underexplored continents like North and South America, the Antarctic, and Africa by mainly Europeans, closed with the beginning of World War I and a shift toward a technological revolution in transportation—airplanes and rockets. The mood of exploration following the Great War remained consistent in terms of the themes of accomplishment, overcoming difficult environments, and nationalism. Human space exploration was simply a Cold War extension of this period and those goals.

The American West

Literature on visual rhetoric, the persuasiveness of images, helps define the ways in which these rationales become relevant during projects that visually document unknown places. While traditional artists accompanied many early expeditions to the American West, photography made visual representations even more useful to project leaders as a means of reinforcing the utility of the expeditions. While not necessarily garnering the same public notoriety as an elegantly painted vista, photographs became a means of quickly gauging an audience's almost reflexive reactions to places. Technology allowed expedition leaders to turn positive feedback into support for their next project. Unlike text-based arguments, visuals tend to prompt a comprehensive, quick, and visceral response by viewers.[8] Exploration photography, therefore, became an efficient way to spread a visual rhetoric that reinforced textual rationales circulated prior to each expedition season. Charles Hill's insightful comment that "Rhetorical images are ubiquitous, powerful, and important" connects the cultural significance of exploration photographs to the intensity of their resonance in public memory.[9]

Exploration of the trans-Mississippi West came on the heels of the formation of the United States itself. The increased desire to explore territories at this time indicates a transition in the relationship between regional and national culture.[10] As the nation began to expand, largely through the Louisiana Purchase, the regionally focused population needed to adjust to a nationalized perspective. Rhetorically, the idea of this natural expansion of the nation to encompass the land mass from the Atlantic to Pacific Ocean was consolidated in the Jacksonian era under the term *Manifest Destiny*. Therefore, many viewed territorial expansion and the exploration of those lands as a natural process, despite one obvious problem: the inhabitants already living there. In fact, art historians discuss William Henry Jackson's landscape photographs as the definitive expressions of Manifest Destiny, the ultimate method for creating a mythological image of a place.[11] If the images were to encourage westward commercial development and habitation, an apparent lack of residents lent itself to the idea of the West as a blank canvas of sorts.

To gain support through claims of advancing scientific knowledge and economic expansion, survey leaders appealed to other respected members of the academic community to write to Congress on their behalf. In an 1870 letter to Representative James A. Garfield, Smithsonian Institution Secretary Joseph Henry requested passage of funding of John Wesley Powell's second Colorado River expedition because: "The region through which the river and its branches flow is one of the most remarkable on our continent. . . . This region is, therefore, highly interesting in a scientific point of view and also in that of the economical application of a portion of it to agricultural purposes."[12] Henry, familiar with Powell's earlier expedition and photography from the region, saw the Smithsonian and Powell's research goals as convergent, evidenced in Powell's later role as director of the Smithsonian's Bureau of Ethnography until his death in 1902.

As Americans debated the political and racial character of added territories in the mid-nineteenth century, the government expanded funding for expeditions. While early expeditions were mainly military in origins and character, support for them waned with the onset of the Civil War in the early 1860s. The Civil War gave prospective expedition leaders a chance to rationalize the need for additional expenditures, this time to include scientific investigation as not a secondary but primary purpose. The interlude of the war gave surveyors time to refine the case for their work, and a chance for camera technology to mature and become both more portable and easily reproduced.[13] Attention then turned to the desire for western expansion supported by a public highly receptive to photographs other than those of war.[14] Exploration images came to symbolize a future full of peaceful expansion on multiple fronts. During Powell's excursions, "the West's unique landscape, as it was revealed in the art of the survey artists, forced the scientific mind into new depths of geological time and inspired a new respect for the natural process."[15] Surveys and their accompanying documentary images (both impressionistic and literal) enlarged the concept and mythology of the West and provided commercial and intellectual rationales for additional government support.

The rhetoric of frontier promises, opportunities, and natural wonder allowed American Western expedition leaders and organizers to accumulate the necessary financial support for their exploration. Concurrent with these mid-century expeditions was the development of a process for reproducing scenes and permanently printing them on a surface, most famously perfected by Louis Daguerre of France. His development, and those of others working on the photographic reproduction process such as Englishman Fox Talbot, began a process of what Walter Benjamin called the democratization of the image during the age of mechanical reproduction.[16] No longer were an artist's impressions required to represent a real scene: a machine could bring those images to anyone able to pay for it. During the earliest period of photography, primarily the 1840s, the technology of producing photographs was simply too complicated to document events in the field and too fragile to transport from place to place.[17] Once the technology caught up to the needs of those wishing to depict the outdoors, representing the discoveries of western expeditions became possible, at least in theory.

Timothy O'Sullivan, the famed photographer who trained during the Civil War with noted photographers Mathew Brady and Alexander Gardner, served on multiple U.S. government expeditions, including two to the American West. While working for geologist Clarence King on the Geological Exploration of the Fortieth Parallel from 1867 to 1869, O'Sullivan created photographs that encapsulated his ideas about the landscape while also conforming to King's narrative about the geology of the territory. His iconic image of the Carson Desert of Nevada showing an ambulance carriage pulled by a team of horses may well be one of the most reproduced photographs of this period (Figure 3.2). This image embodies important themes visible in photographic work for the near-century of exploration that would follow.

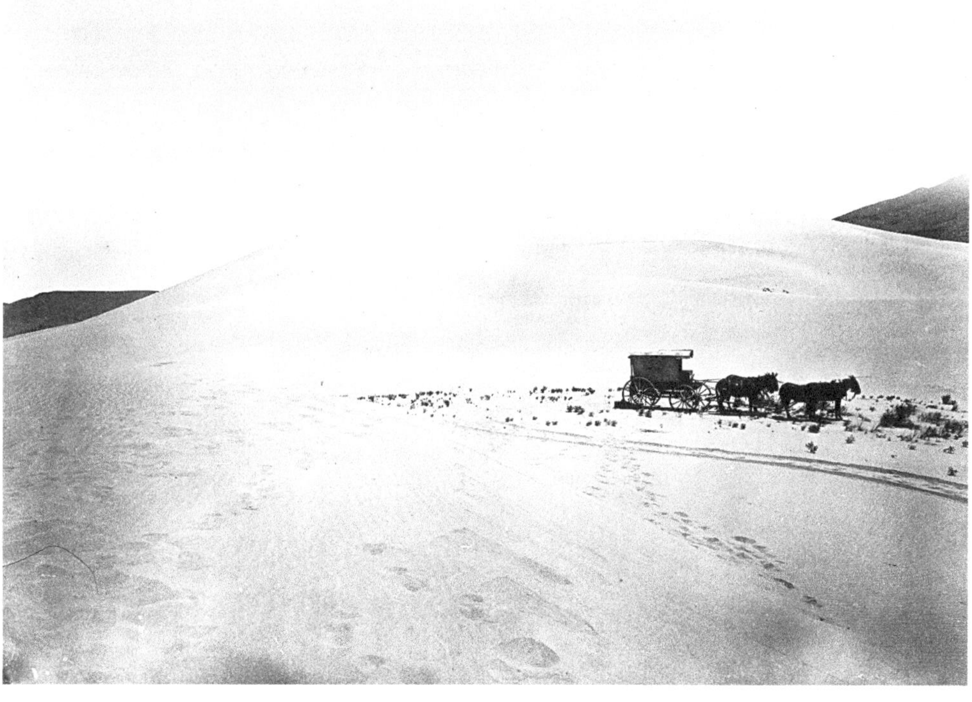

Figure 3.2. Timothy O'Sullivan's ambulance wagon and portable darkroom used during the King Survey rolls across the sand dunes of Carson Desert, Nevada, 1867, 77-KS-346O (National Archives).

O'Sullivan's photograph of a horse-drawn covered wagon traversing a desolate and sandy landscape expresses just some of the hardships encountered by early explorers of the American West. The wind-contoured foreground sand, dotted with footprints and wagon tracks, draws attention to the sparse mid-ground vegetation and four-horse team pulling a recently turned wagon. Behind the wagon, sandy dunes and perhaps rockier hills beyond shape the horizon against an empty sky. Subtle clues indicate the staged nature of the scene, making this image less about capturing a journey in process and more of a reflection on the hardships of exploration. The sweep of the wagon tracks shows how the wagon moved from the lower right and circled back to its final position for the photograph. The one or possibly two sets of human tracks in the sand indicate at least one trip of the photographer or an assistant to and from the wagon. These elements indicate O'Sullivan's consideration for arrangement and framing the scene, expressing a message of the West as arduous for the expedition team, or at least that wagon.

O'Sullivan's depiction of known objects such as the horse team and wagon physically situated humans in the context of that space without showing actual people. While not explicitly about the expedition itself, O'Sullivan's image does show the means of transportation of people to and through the location explored. He wanted the viewer to know how people got to this remote location, yet another key component of visual narratives of exploration. This photograph reflects on the grandeur of the landscape and the magnitude of human interaction with the landscape. The composition places familiar and scalable objects in the frame to give viewers a point of reference for the size of the desolate hills. Indications of the scientific aspects of the expeditions are more difficult to decipher from this particular image, but the wagon itself and the passage of humans through unfamiliar territory allude to discovery and the collection of information in an abstract sense. The meaning of O'Sullivan's photograph is essentially that this so-called untouched landscape was available and accessible by the kinds of implements seen in the foreground.

Evolving imaging technology and a public fascinated with the seeming truth in photographs encouraged visual representation of the promise of westward expansion.[18] Expedition leaders of the four Great Surveys of the West—Clarence King, George Wheeler, Ferdinand V. Hayden, and John Wesley Powell—made the case to their financial sponsors that visual representations were critical to successful expeditions as promotional tools in support of further scientific and economic development. To provide the greatest range of visual information, expeditions required photographers, artists, and topographic drafters. Photographic support of these expeditions by professionals such as John Hillers, William Henry Jackson, and Timothy O'Sullivan, and the continuing mission of the United States Geological Survey (USGS) to document the lands of the United States, ascribed a certain purpose onto the photographs. These images served primarily as evidence not necessarily requiring aesthetic appeal. They also provided expedition sponsors with an expanded physical view of the nation itself, making them crucial for developing government and public support for additional exploration. Images are excellent descriptive tools but do little to explain an event or provide substantive details.[19] Survey leaders, politicians, and scientists therefore required visual resources and accompanying text to generate interest. Images and text worked together as rhetorical devices for creating an idea of the American West and a rich cultural interpretation of that place seen in films, comic books, and thousands of other media types over a century and a half.

Scientific, commercial, and military efforts pushed native peoples farther West and made use of textual and visual documentation from expeditions. Programs continued to expand the catalogue of plants and animals of the region and began to examine the geological resources that would spur the government and mining companies to sending their own groups to document the land itself, continually encroaching on the territorial claims of native peoples. Notable at this time were the mapping of the Oregon Trail area and parts of California and Oregon by John C. Fremont and the U.S. Army Corps of Topographical Engineers in the 1840s, and the commercial efforts to find a reasonable passage for a transcontinental train route through the 1850s.[20] While not a subject of this study, we cannot ignore the ways in which the photographic surveys of the American West played a significant role in representing inhabitants to non-natives, serving only the needs of western expansion.

During exploration of the American West, project leaders hired photographers with a wide range of training and experience. From the on-the-job training of John Hillers during Powell's second Colorado trip to self-taught photographer William Henry Jackson, primary image makers of western expeditions arrived on teams with little formal photography education. Initially, Powell hired Hillers as a boatman who occasionally assisted the expedition photographers, E. O. Beaman and James Fennemore. He later replaced both of them. Even though he acquired his skills along the road, Hillers became not only a talented photographer, but also a fine artist.[21] These men defined the role of an artist-photographer on an expedition, meaning that they determined a specific point of view, the scene content, what camera to use, the settings required, how to process the film, and then played some role in the use of that image as part of a narrative of the expedition.

Timothy O'Sullivan and William Henry Jackson entered survey work from a different direction than John Hillers. Both spent years during the Civil War either photographing or sketching scenes of battle, O'Sullivan as an employee of Washington, D.C.-based photographer Alexander Gardner, and Jackson as an artist-observer. O'Sullivan unfortunately left little textual evidence of his point of view or thought process in framing a scene as he did not keep journals. The photographs Gardner included in his 1866 book, credited to O'Sullivan, are indicators of the style O'Sullivan carried through his survey work. O'Sullivan's work was often collaborative in terms of determining certain shots, showing he understood the audience of geologists, surveyors, and others in need of details about the western environment.[22] The conjunction between artist, photographer, explorer, and scientist became a crucial element in these expeditions, and later ones as well.

Similarly, Ferdinand Hayden and William Henry Jackson developed a working relationship that provided each with assurance of professionalism. Hayden's ego as a surveyor melded with Jackson's vision of portraying the dramatic landscape and the heroism of the survey teams.[23] Jackson's background working for the Union Pacific Railroad and as a photographer in Omaha prepared him for survey work as an artist. Jackson had already bought into a somewhat fanciful concept of the West: his visualization of the land through photographs would continue to support an idealized place.[24] His images could in essence sell the concept of westward expansion, and Hayden saw the great potential in using Jackson to further his own desire to perpetuate his expeditionary work.

The Antarctic

Exploring the Antarctic continent required perhaps even more rhetorical and financial backing than movement westward across the American landscape. Subsidies needed to cover not only basic travel expenses and supplies, but also an entire ship. Like what came before and since, rhetoric for Antarctic exploration focused on potential scientific discoveries, economic benefits, and a need to succeed on a global level. The potential of a British expedition reaching the South Pole before the Japanese, Americans, Germans, or Norwegians most certainly counted a great deal toward gaining government and academic support when Victorian Britain's colonial success began to wane. Robert Scott's expeditions had explicit nationalistic and imperialistic overtones at the height of the British Empire's power, though such desires could arise from different sources of power or needs. Susan Solomon explains that a sort of "Antarctic fever" developed as part of the repercussions of the 1909 controversy between Peary and Cook about who reached the North Pole first. That excitement inspired fundraising efforts around the entire empire: eight thousand volunteers applied to participate in the expedition. As the "official" British expedition to the South Pole, Scott's status as a national hero was secure, no matter the result.

Ernest Shackleton was "a man of romantic ambitions" and made a firm commitment to finding a place of honor for himself among the greatest explorers of his time.[25] Raised in a middle-class Irish home, he had no means of self-supporting his ambitions but married well. By cobbling together enough financial support from wealthy investors, Shackleton attempted his first polar trek in 1907, coming within one hundred miles of the South Pole in 1908 before poor weather, a lack of supplies, and loss of their final pack pony forced the team to return to their ship, the *Nimrod*. Despite the failure, the British heralded his accomplishment, calling him a national hero, and Shackleton received a knighthood from King Edward VII in 1909.

Frank Hurley photographed HMS *Endurance* trapped in an ice floe during the monotonous days of Shackleton's Weddell Sea expedition in early 1915. Images such as the ship seen at night (Figure 3.3) are iconic and representative of the extreme challenges of Antarctic exploration. Hurley, a self-made professional photographer from Australia, joined Shackleton as not only expedition photographer, but also a full-fledged crewmember with duties commensurate with his availability and skills. Hurley managed to capture over five hundred glass plate negatives during the expedition, many of which were lost when *Endurance* was abandoned to the voracious ice in late 1915. Thanks to Hurley's photographic skills, some images of the voyage survived to satisfy a fascination with human exploration of an almost-alien landscape, and perhaps served as a warning of the failures possible in those places.

Like Timothy O'Sullivan's photograph of the ambulance wagon traversing the Carson Desert, Hurley captured a sense of human scale in Figure 3.3 by placing the expedition's sailing vessel at some distance. This gives the viewer a sense of how ships traveled to this frigid and dangerous land and the varied dangers posed by

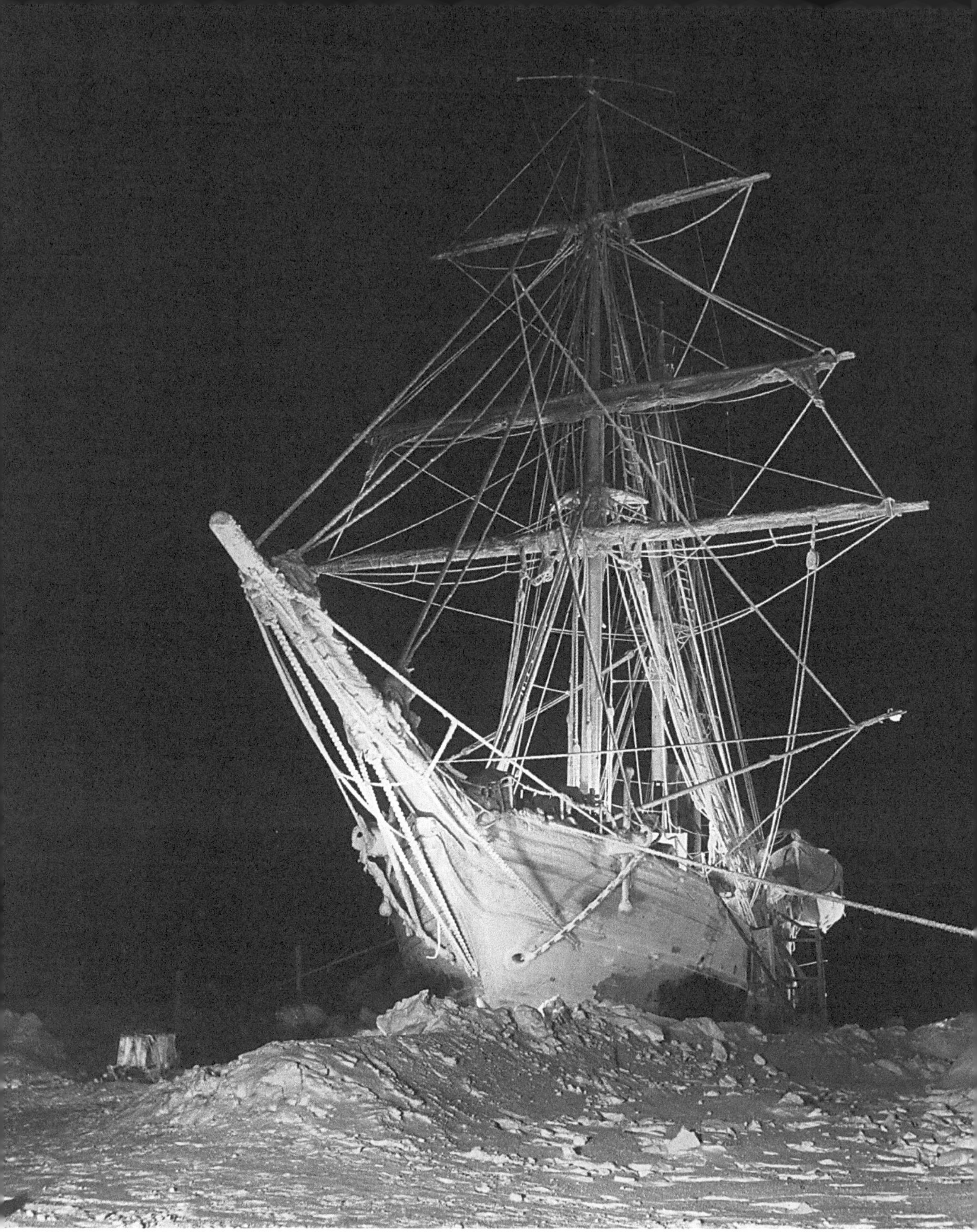

Figure 3.3. *Endurance* at midwinter, photographed by Frank Hurley, 1915, P66/18/43 (Scott Polar Research Institute, University of Cambridge).

the immense, ever-changing, icy landscape. Hurley contextualizes ice as not only different from what most of the audience would find familiar, but also as the most significant feature of the Antarctic. An imposing view of the ship in the scene gives an almost elongated view of the single human element and its battle against the ice. And like O'Sullivan's subtle cues in his image of the ambulance wagon, Hurley's inclusion of the transportation method on a scientific voyage (the ship) and the space between it and the photographer alludes to the sense of a great swath of ice to explore, study, and document. The contrast between foreground features Hurley referred to in his caption as ice flowers and the background of pressure ridges formed by the ice floe contributes to a sense of difference that would presumably intrigue those who studied ice formations. Hurley's still photograph brought together the key components seen across exploration photography that exemplify the goals expedition leaders had for using cameras to create a visual rhetoric of the project.

In the case of Antarctic trips made by Robert Falcon Scott in 1910 and Ernest Shackleton's cross-continent expedition in 1914, themes of transportation and discovery are evident in both the attempts and the repercussions of their failures. By the time of Antarctic expeditions of the early twentieth century, photographs were not only commonplace in news reporting, but also expected as a part of heroic journeys to remote lands. These experiences connect visually and culturally to exploration of the American West, sharing what is described as the power of images residing in their perceived ability to frame an event and suggest universal values attached to that event in public imagination.[26] Despite successes and failures, explorers provided heroic visual narratives, both mythological and culturally relevant to form a public perception of a place. The American West, on the one hand, became conquerable and a place to populate and cultivate in part because others were already doing so. Antarctica, on the other hand, with its treacherous, icy, difficult-to-inhabit landscape, offers a far better analogy for the experience of astronauts. The Arctic also became significant in American culture because it generated a "widespread belief that exploration is an impulsive and instinctive activity, deeply rooted within the human psyche," which folded into a larger "pandemic of exploration" in the early twentieth century that included exploration of the Antarctic.[27] As both Scott and Shackleton spent considerable effort to ensure a visual record of their projects survived, it seems fair to state that they provided a more substantive and reliable visual narrative by which Britain and the world could understand the rigors of polar exploration.

The physical and intellectual requirements for taking photographs in the Antarctic were complicated for two of the most celebrated and remembered expedition photographers of their day. Frigid temperatures, ice floes, and charismatic expedition leaders challenged photographers like Herbert Ponting, the scientific photographer on Robert Scott's disastrous last expedition on the *Terra Nova* (1910–13), and Frank Hurley, who accompanied Ernest Shackleton on his ill-fated journey on the *Endurance* (1914–17). As a member of the scientific team, Ponting never traveled far from the coast or nearby depot points set up to support the polar team. He therefore needed to train Scott and other expedition members for photographing the remainder of their trek. Searchers found their photographs and camera not far from the frozen bodies of Scott and his team in 1912. Ponting's photographs documented the crew's experiences, wildlife, and the weather, but had little to do with anything other than science and information. Following the expedition's tragic end, Ponting took it upon himself to turn his photographs into a means of honoring those who died. He gave public and private lectures and printing albums for the families of the expedition members, supporters, and the government. His account of the expedition, however, took ten years to compile and relied heavily on Scott's journals, rather than Ponting's own photographs.

Shackleton hired the Australian Hurley for the journey aboard the *Endurance* to capture photographs during extreme conditions and to chronicle everything: the treacherous nature of the pack ice, life aboard the ship, and especially the expedition's unexpected struggle to escape a sinking ship and survive on a desolate continent. During the voyage, the crew experienced temperatures as low as -34° C in the depths of winter, rode out numerous blizzards, and faced a life of diminishing supplies.

Thanks to the importance placed on image making by Shackleton and his patrons, the ship was spacious enough to accommodate a darkroom and image storage area for Hurley. He secretly salvaged the plates he created from the flooded, ice-locked ship, despite Shackleton's opinion that the plates were worthless. The two eventually agreed to a deal whereby Hurley could keep 120 developed plates, a small Kodak camera, and three untouched rolls of film to document the rest of the trip. Moreover, while the voyagers went cold and hungry for weeks on end, eventually sacrificing their sled dogs to avoid having to feed them precious rations, they understood their lives and legacy appeared on Hurley's glass plates.

Exploring the frigid Antarctic with a camera threw up obstacles up for photographers like Ponting and Hurley. Their prior experiences of travel and exploration photography, however, prepared them for the environment and demanding leaders.

Ponting joined the *Terra Nova* expedition based solely on Scott's reputation and despite a prior work commitment.[28] He developed considerable experience over the prior ten years doing freelance photojournalism for nature periodicals, and his 1910 publication on Japan earned him a Royal Geographical Society (RGS) fellowship.[29] With that connection and a reputation for producing elegant narratives through visuals, Ponting became a significant addition to Scott's voyage despite not making the final trek to the Pole. For the final leg of the trip, Scott requested that Ponting train team members to handle a small camera themselves (Figure 3.4). Scott, keenly aware of the ongoing controversy between Robert Peary and Frederick Cook over reaching the North Pole, understood that having photographic proof and accurate navigational evidence of the accomplishment meant the difference between instant recognition and battling naysayers or competing claimants.[30]

Perhaps more than any of the photographers, Frank Hurley acquired actual photographic and technical training through schooling. He attended a Sydney technical school and purchased his first camera at age seventeen. Alasdair McGregor, Hurley's biographer, depicts the photographer as motivated (he opened his own postcard business as a young professional), adventurous, and willing to take risks to

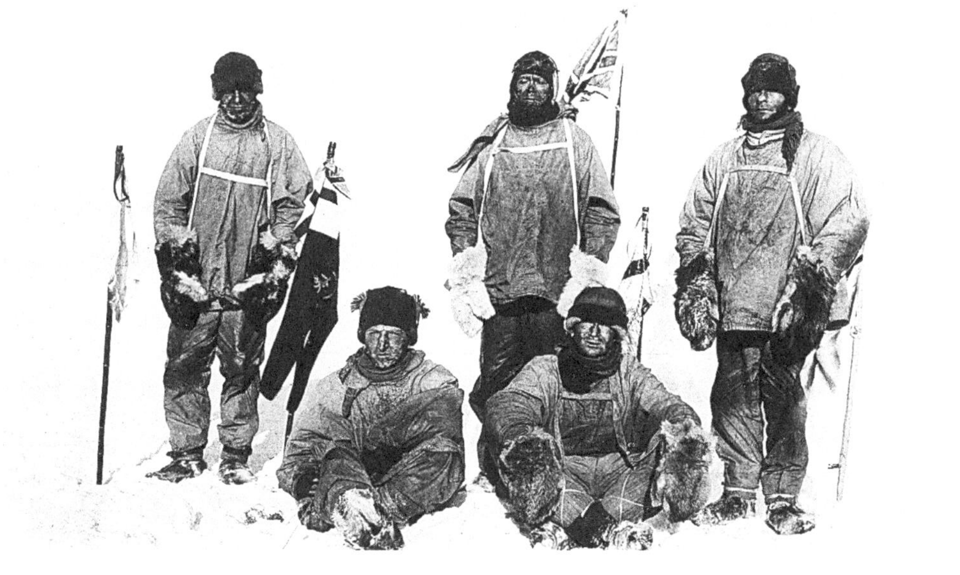

Figure 3.4. The Polar Party at the South Pole. Scott, Oates, Evans stands, Bowers and Wilson sit in front with the Union Jack flag on a pole behind, photographed by Henry Bowers, 1912, P2005/5/1346 (Scott Polar Research Institute, University of Cambridge).

get the best image.[31] Hurley was also an innovator, using the popular and recently developed Paget color process during his work in Antarctica. This technique made the glass plates very sensitive and, therefore, advantageous in the low light conditions of the polar environment. Such was Hurley's dedication to his craft that Shackleton smashed the glass plates deemed more than the team could salvage and transport, so Hurley would not be tempted to save them and overburden the team.

During the heroic age of polar exploration, Shackleton proved to have an innate ability to exploit British nationalism and imperial rhetoric in the funding of his expeditions at the height of the British Empire. The journey's official name, the Imperial Trans-Antarctic Expedition, indicated its conception as a grand and inspirational effort.[32] Shackleton's success in convincing his own government and wealthy investors, using other creative means of raising money, serves as a testament to his skill in using rhetoric for self-promotion. He appealed on nationalistic grounds and encouraged investment from royal societies, wealthy individuals, and even public schools. What aided him most was the willingness of the press to sing his praises.[33] To make up the balance of his funding needs, Shackleton also sold his services as a lecturer and national hero. By offering publication rights to his story, both written and visual, Shackleton leveraged every resource available to accumulate the funding necessary to make his dreams come true in 1914.

Perhaps even more than explorers of the American West, Ernest Shackleton found heroic, nationalistic, and visual rhetoric necessary to interest a wide range of investors in his venture to cross Antarctica, just as NASA would use images to appeal to Congress and other audiences for funding. In an incredibly confident letter to Winston Churchill, then head of the Royal Navy, Shackleton attempted to enlist the support of the government by comparing his own risks for science and country side-by-side with the risks taken daily by Churchill and the Navy. He explained the relative ease of the voyage and benefits to exploring the Antarctic compared to endeavors such as military aviation or even crossing a busy city street, attempting to reinforce the value of supporting his effort.[34] Shackleton did not stop at the Navy. For its part, the Royal Geographical Society, no stranger to funding exploration in the interests of the Empire, begrudgingly granted Shackleton £1,000. They hoped to access the scientific data and the photography expected to return: they were less than pleased with the returns from his earlier attempt to reach the South Pole. Shackleton, by design a private explorer and not officially directed by his country's government, needed to sell his exploration plans to potential supporters, audiences for the information he would obtain in similar terms to how NASA needed to garner support for human spaceflight. In the climate of a world on the brink of a world war, he had difficulty convincing wary investors to ignore their instincts and provide money toward his ambitious plan.[35]

Shackleton capitalized on the increasing use of camera technology at the time.[36] Seeing its success on the Scott expedition (1910–13), Shackleton expected to tell his story in similar visual terms, though with a less disastrous outcome. Until recent decades, however, iconic images captured by Frank Hurley of the *Endurance* experience hardly competed with those of the Scott expedition taken by Herbert Ponting. Perhaps visuals of a heroic dead explorer meant more to the public than a heroic living one. It was actually Scott that first identified the necessity of photography after the expedition to pay off financial debts.[37] After hearing of Scott's death, Ponting capitalized for years on his photographs, as had photographers on the great surveys of the American West. Shackleton took advantage of the wisdom of Scott and marketing of Ponting to prepare a photographic plan for his expedition in 1914.[38]

EXPLORING SPACE WITH A CAMERA

Expedition rhetoric, experiences, and images from the American West and Antarctica set up participants and viewer expectations for how we would see and remember space exploration. Photographs became a means of making a positive impression on constituents about the value of those experiences and imprinted them on our collective cultural memory. Astronaut photography made a strong visual contribution to how we continue to see the 1960s, the Space Race, and the Cold War, not only because the images were spectacular, but also because of tremendous efforts that made those previous expeditions possible. A closer look of how astronaut photography contributed to American visual culture requires a three-part examination.

For space exploration, rhetorical links with westward continental expansion inherently yielded powerful and similar visual representations. These connections demonstrate how NASA used both words and images to garner public and political support throughout its existence. Frontier rhetoric used by NASA and its supporters conveyed a notion "of the American nation, [as] a triumphant and glorious story of success, with the complex stories of Indian conquest and African American slavery simply ignored and eliminated."[39] While NASA may not have fully considered the implication of using frontier rhetoric, "The notion of a heroic journey accomplished was one that could be appreciated, even shared, by the potential audiences of the survey photographs,"[40] and "the promoters of space exploration and development may well qualify as the nation's most committed and persistent users of the frontier analogy."[41] Skilled NASA communications and political efforts exploited nostalgic feelings of an imagined and idealized American past by harkening back to supposed benefits of western expansion by using the exact same rhetoric.[42]

Once most of the Earth's notable geographical features became familiar visually from a ground-level perspective, the next step in human exploration required photographs as a source of evidence and visual rhetoric. Astronaut explorers, trained to operate cameras and frame images for a variety of uses, were entrusted with encapsulating their experiences and the mission's rhetorical goals in a visual medium. NASA designated one Apollo astronaut per mission as having primary responsibility for photography, but the equipment circulated frequently among them or swapped between the moonwalkers during EVA work according to scripted mission plans. For Apollo 11 lunar surface work, mission commander Neil Armstrong, as the first to step on the Moon, also had responsibility for the majority of early photographic work. He shot a full 360-degree panoramic series of photographs soon after his first steps, and then a long series afterward including setting up the American flag. His image of Lunar Module Pilot (LMP) Buzz Aldrin standing next to the flag with the lunar module nearby is an example that contains a set of elements commonly found across exploration photography.

In his first moments on the lunar surface, Aldrin glanced around and described the scene as "magnificent desolation." The gray and brown lunar soil, dotted with small rocks and littered with boot prints, extends from the foreground to a distinct horizon that sharply delineates the blackness of space from the emptiness of the lunar landscape. Set against this stark backdrop are the vibrant white astronaut spacesuit, the American flag, a television cable laying along the ground, and the shadowed legs of the lunar module to the left in the mid-ground. The background is empty other than the hint of rising terrain and the extended lunar module shadow. This barren, desolate scene begs the question of what drew these suited Earthmen to such a dreary place.

In an almost indistinguishable salute to the flag, Aldrin draws our attention to the focal point of the photograph, one rarely seen so explicitly in images of exploration other than as a signifier of the team's ultimate triumph over nature. The presence of the astronaut and the flag in particular set very familiar human elements into an unknown alien environment, with boot prints and a human shape giving scale to the setting. While only a subtle clue for general viewers, the inclusion of the LM to the left helps frame the voyage and the means of transportation to that place. Unlike many existing exploration photographs featuring imposing natural landscapes or those from later lunar missions to more mountainous regions, this image from Apollo 11 gives a sense of grandeur in just the way Aldrin described in his initial assessment: the desolation is what makes the scene sublime. Perhaps even more dramatically though is the knowledge that survival in this harsh environment fully depended on that white space suit and a personal life support system.

None of the scientific instruments deployed on the mission are visible in this photograph, nor are any of the soil and rock sampling tools used by the astronauts. Those components would illuminate the scientific nature of the project, but other clues exist to give viewers an indication of exploration and discovery. Lunar photographs feature small crosshairs (fiduciary lines) on each the frames created by a reseau plate. These lines are used in photogrammetry to precisely measure distance and size when other tools are impractical or unavailable. The landscape itself also presents a great unknown, inviting inquiry made possible by human traffic through the space as indicated by the tracks of the astronaut boots. In summary, this single image includes a wealth of symbols that make it an excellent representation of how NASA exploration photography fits perfectly into the tradition of those captured by Timothy O'Sullivan and Frank Hurley.

READING VISUAL THEMES

During the infancy of photography, traditional artists played a major role in formulating a vision of the American West. The potential to reach viewers effectively meant including intense color, scale, and the embodiment of emotion through the grandeur of the landscape (Figure 3.5). Black-and-white photographs of the time could not reproduce those emotions. Artistic impressions of American western exploration continue to inform perceptions of space in our world today. During planning for interpreting data from the Hubble Space Telescope, scientists looked for inspiration in the work of nineteenth-century painters such as Thomas Moran, the artist who traveled on Ferdinand Hayden's 1871 survey of the Yellowstone region, and Albert Bierstadt, who independently traveled and depicted western scenes.[43] Keying in on the familiar, what they knew to be appealing and comfortable to the public, scientists assigned color in Hubble images to replicate known visual references, making those images more immediately understood (Figure 3.6).

Distant and abstract places in the universe seen by Hubble scientists allowed them to create visual links to paintings, evoking the same type of emotional connection between viewer and visual. Artists like Moran and his compatriots used their impressions of the landscape, not a literal interpretation, to play on the feelings of the viewers and inspire additional interest. Whether viewed in a gallery or through a lithographic print in a publication, artists engaged with their audiences through ideas, not science, using vivid colors to strike at the emotions of the viewers. On the other hand, photography's expected depiction of reality connected it more strongly with expectations for scientific evidence. Audiences believed that seeing a photograph connected them in an unmanipulated way with the location. In the cases of exploration, familiar elements within the photographs assisted in creating an understanding and public memory of those expeditions. Nineteenth-century photography only involved color when added later with colorization techniques. By the time of Antarctic exploration, the autochrome process was just coming into use. In such a generally colorless place though, even Herbert Ponting's earliest experimentation with color imaging had little impact on the public audience (Figure 3.7). NASA would find out, decades later, that planning for color photography on the Moon presented similar problems, pitting scientific and public needs against each other when providing astronauts with film magazines for their photographic assignments in a relatively monotone environment.

Sweeping landscape works by survey painters, while rarely reproduced or seen by large public audiences when created, gave those responsible for creating the rhetoric supporting westward expansion—politicians, writers, historians, and philosophers—idyllic images from which their words could suggest a new and idealistic path westward. Angela Miller points out that these visuals created a scenario where "rhetoric that linked nationalism to the geographical unity, breadth, and scale of the New World must be weighed against the evidence of landscape paintings themselves project to modern eyes a view of nature both bounded and contained."[44] The dissemination of photographs of these locations, similar to the artwork in its use as a means of transmitting messages, allowed the positive messages to circulate quickly among the public thanks to the ease of reproduction.

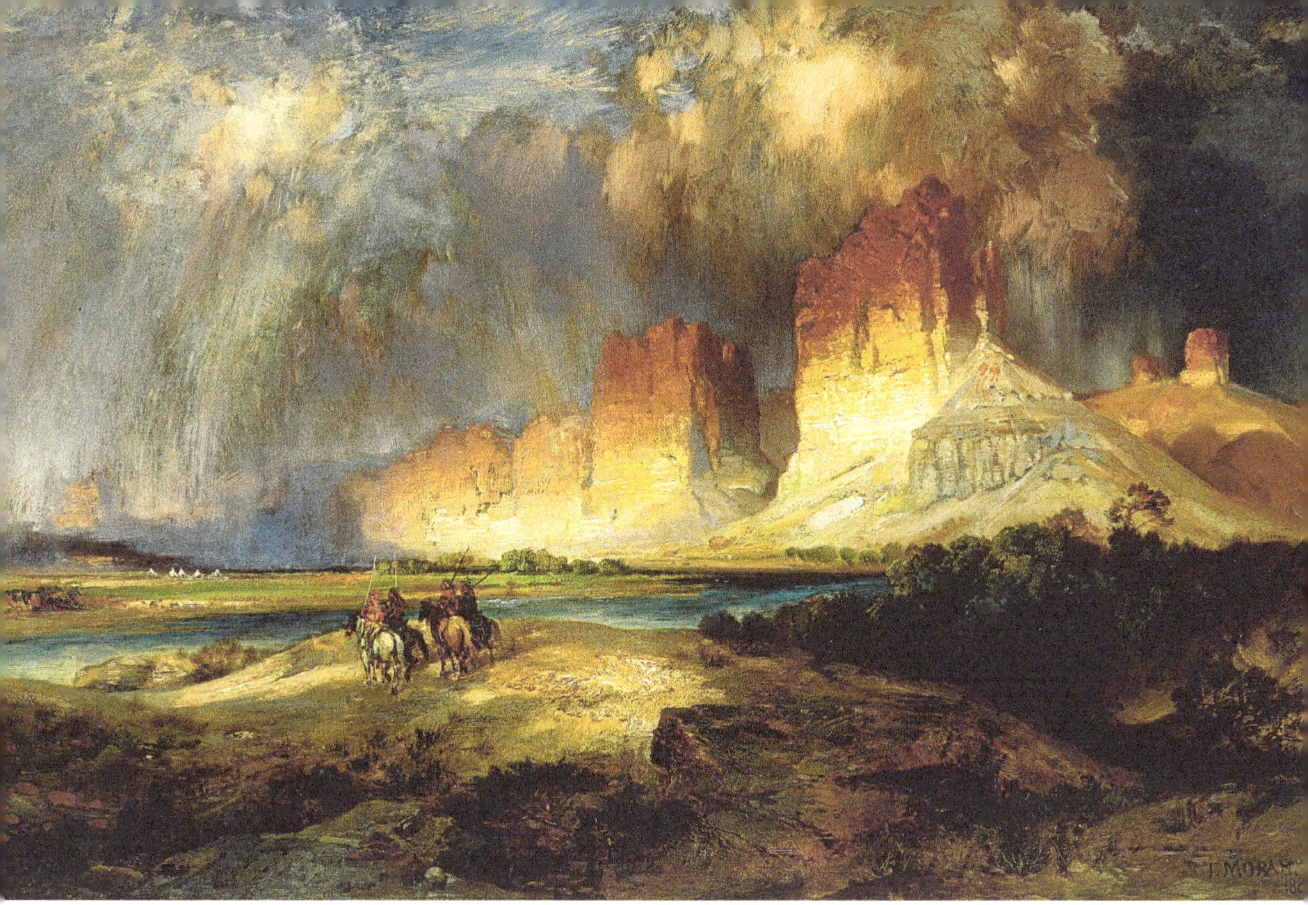

Figure 3.5. *Cliffs of the Upper Colorado River, Wyoming Territory*, by Thomas Moran, 1882 (Smithsonian American Art Museum, Bequest of Henry Ward Ranger through the National Academy of Design).

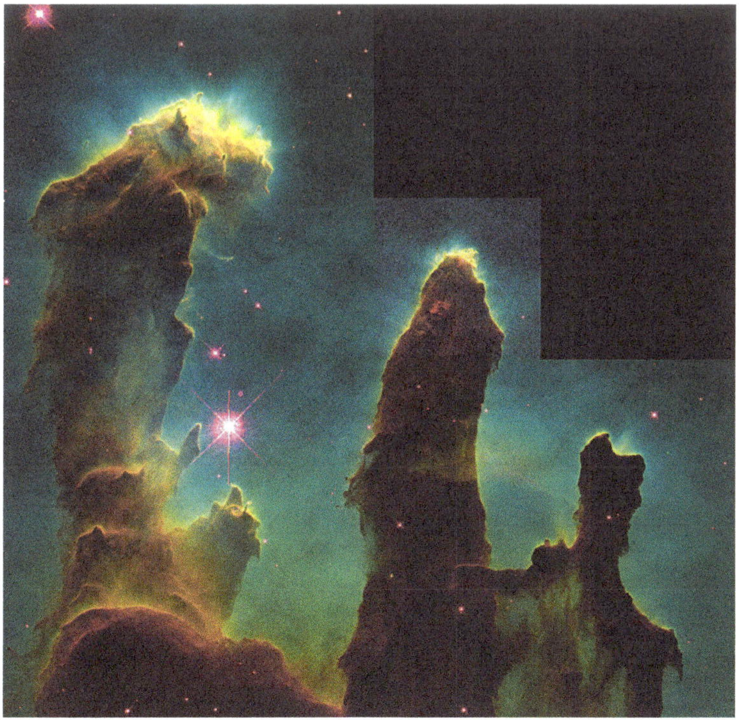

Figure 3.6. *Gas Pillars in the Eagle Nebula (M16): Pillars of Creation in a Star Forming Region*, 1995 (NASA, ESA, STScI, J. Hester and P. Scowen, Arizona State University).

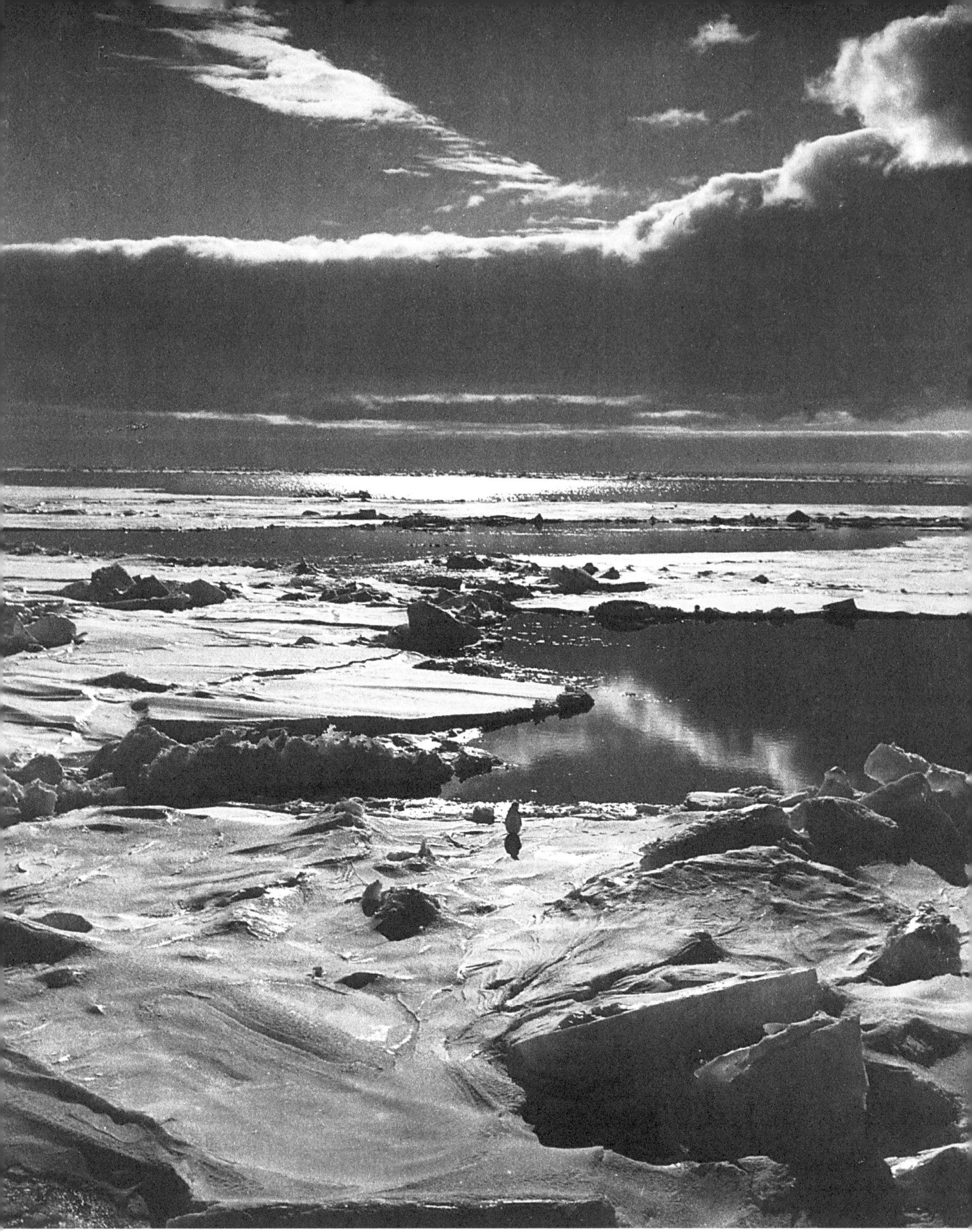

Figure 3.7. Sun over the Barne Glacier, autochrome photograph by Herbert Ponting, 1912 (© Royal Geographical Society [with IBG]).

Frontier rhetoric played a significant role in defining the expectations and outcomes of exploration of the American West, Antarctica, and space, and images of outer space conveyed a desire to conquer in the same sense as many Americans saw the western landscape in the middle to late nineteenth century. Depicting the so-called frontier of America, according to David Wrobel, "has become a metaphor for promise, progress, and ingenuity."[45] Such nationalistic rhetoric flows freely without clear definition of the scope of expansion defined or limited. Visual and textual products, such as artwork, photographs, diaries, and other representations of the landscape, contributed to reinforcing rhetoric and became fundamental to defining a frontier ideologically. Using such rhetoric too freely, however, meant that "The term 'frontier' blurs the fact of conquest and throws a veil over the similarities between the story of American westward expansion and the planetary story of the expansion of European empires."[46] While the term itself wields significant power, promoting a set of American ideas about progress, the use of "frontier" cannot be underestimated in its utility in the context of exploration.

Representative images of exploration almost singularly manage to encapsulate the rhetoric, narrative, and meaning of the projects in ways impossible with words alone. Photographs as fixed images, moments in time, can imprint their messages on viewers through symbolic and literal codes, and their impressions last more readily than those of moving images through the ease of their reproduction, reuse, and transfiguration into new forms.[47] These still photographs became a means for permanently remembering, celebrating, and representing exploration. The content of representative images forms the basis of an exploration project's visual narrative, with hundreds or thousands of other images meant to support that narrative.

The visual continuity of images is striking when examined across human exploration projects over the last century and a half. Some trends are common in photography, many often seen in tourist photographs. Each journey needs an "establishing" shot documenting arrival at the destination. The photographer lays the grandeur of the landscape before the eyes of the audience, inviting varied interpretations but encouraging awe and wonder at the spectacle of nature, what photography literature refers to as the view type of landscape photograph.[48] Over time, such view landscapes become familiar to audiences. Particularly explicit visual links correspond to the concepts of sublimity and technology in American culture seen since the early nineteenth century's industrial revolution.[49] While grand paintings of the American West by Thomas Moran inspired a sense of awe among viewers, technology made reproducing photographic representations of the natural sublime easy and accessible to a wider audience.

Other than the view type of landscape photograph, other visual trends permeate exploration photographs. These amazing adventures, laden with frontier or pioneer rhetoric, typically drew on the sense of the far-off horizon to bracket the landscape within the bounds of the camera and the sky, expressing a vastness of the place, but also encouraging viewers to ponder what lies beyond that line in the distance. Creating a vanishing point became a visual cue to direct the gaze to a distant point. As the ground converges with the sky at the horizon line, drawing the eye and mind toward the distance and what lies beyond that line (Figures 3.8, 3.9, and 3.10). The land between the viewer and the horizon begs explanation. What is seen is just as important as how it is seen, giving those who control the creation and distribution of the images a link to the rhetoric that sold the effort in the first place.

Images became the means by which explorers justified their research, just as NASA continues to use *Earthrise* to promote the accomplishments of the space program. The Powell survey photographs by John K. Hillers played a significant role in the expedition in that "The artistry of these pictures supported their social and political functions: these views could at once persuade, instruct, impress, and delight."[50] Side-by-side comparisons between landscape photographs from the West

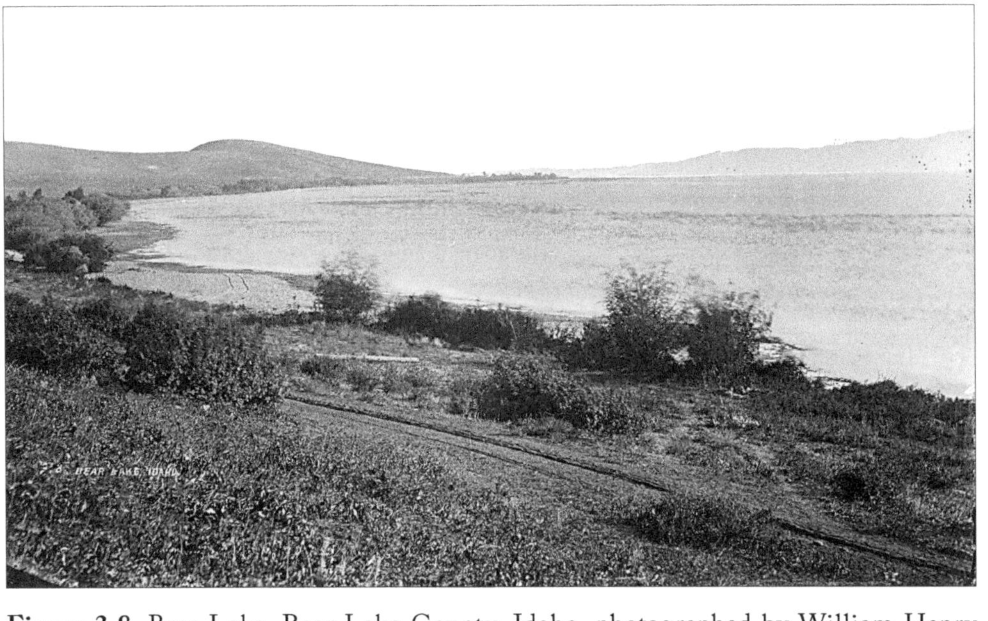

Figure 3.8. Bear Lake, Bear Lake County, Idaho, photographed by William Henry Jackson, 1871, 57-HS-124 (National Archives).

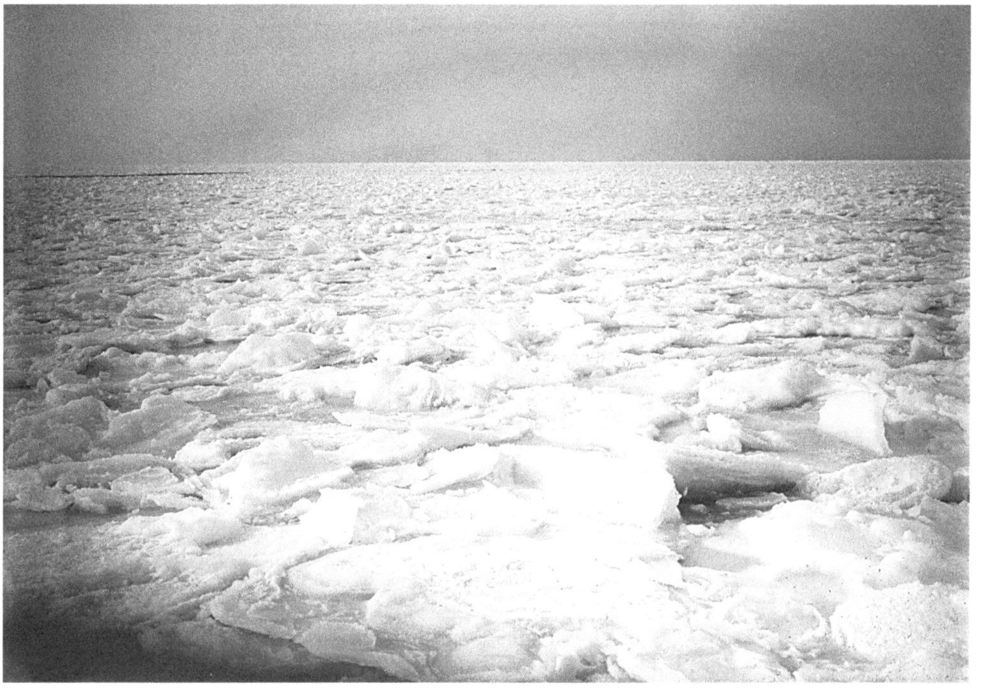

Figure 3.9. Heavy pancake ice in which the *Terra Nova* was held up while trying to rescue the Northern Party, photographed by Herbert Ponting, 1912 (Scott Polar Research Institute, University of Cambridge).

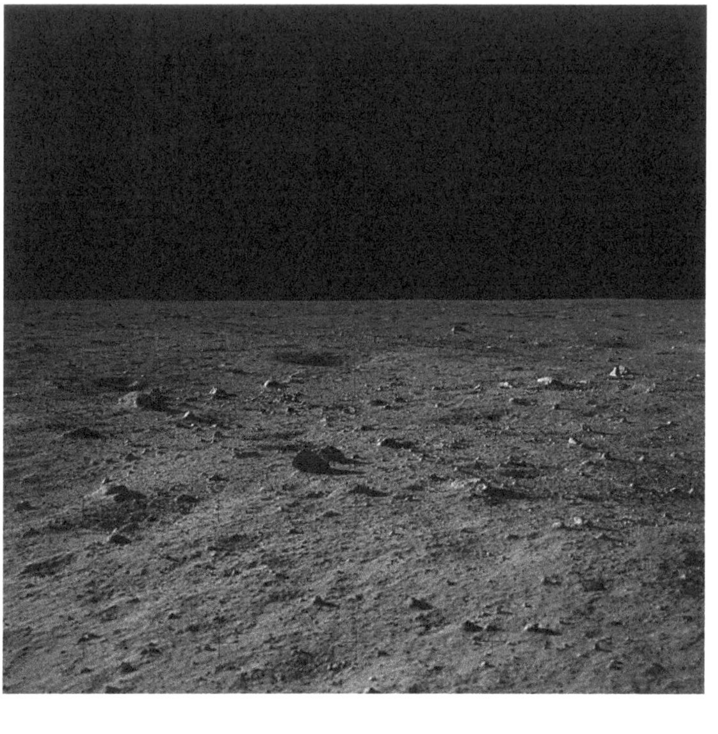

Figure 3.10. Second photograph captured by Neil Armstrong after stepping onto the lunar surface during the Apollo 11 mission, July 20, 1969, AS11-40-5851 (NASA).

or Antarctica and *Earthrise* reinforces rhetorical continuity (Figures 3.11 and 3.12). These stark and uninhabited scenes mirror the stark, rocky, uninhabited landscape of the lunar surface (Figure 3.13). In all three cases, the horizon and composition lead the gaze toward the vanishing point. Horizons hint at the idea of the great beyond, a future, and the content of the image aligns in a clear path between viewer and the potential that lies beyond that line. Scholarship on American West photography offers further insight into how scholars could analyze later exploration photography as well.

By analyzing images in context, linkages emerge between exploring a new region, conceptions of that place, and changing pictorial representations of it. Photography becomes a means of connecting its uses, meaning, and changing cultural ideas of places.[51] Seeing the Moon in 1969 meant different things than it does to scientists using the Lunar Reconnaissance Orbiter today. Not unlike travel by wagon and horse to the American West or by ship to the Antarctic, space presented serious challenges to transporting camera equipment. Americans

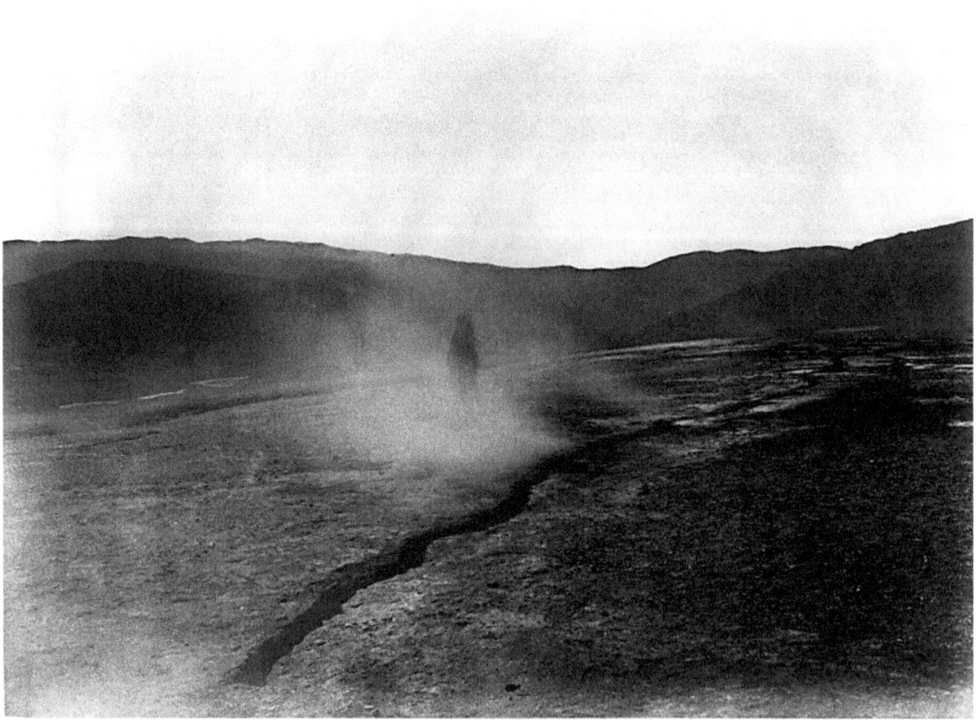

Figure 3.11. Steamboat Springs, Nevada, photographed by Timothy H. O'Sullivan, 1867, LC-USZ62-74452 (Prints & Photographs Division, Library of Congress).

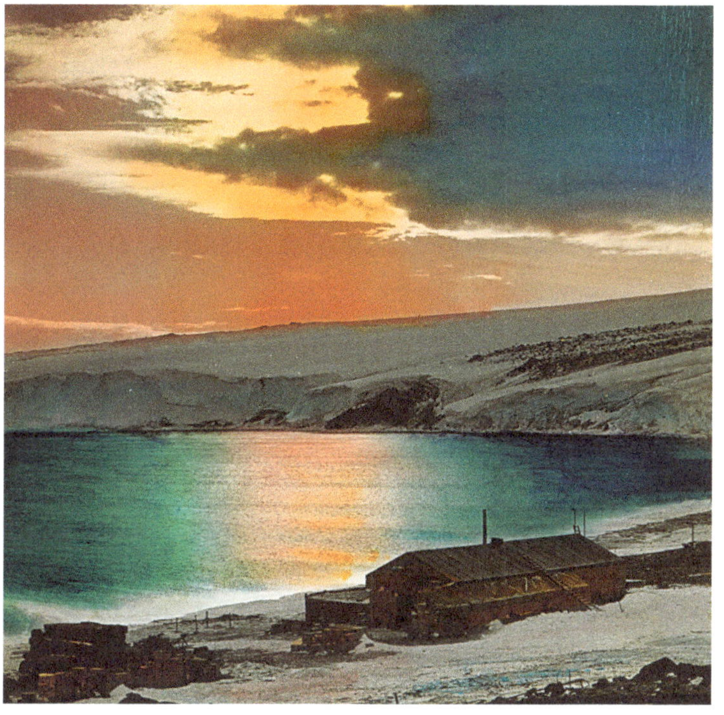

Figure 3.12. Christmas Eve in the pack, photographed by Herbert Ponting, December 24, 1910 (© Royal Geographical Society [with IBG]).

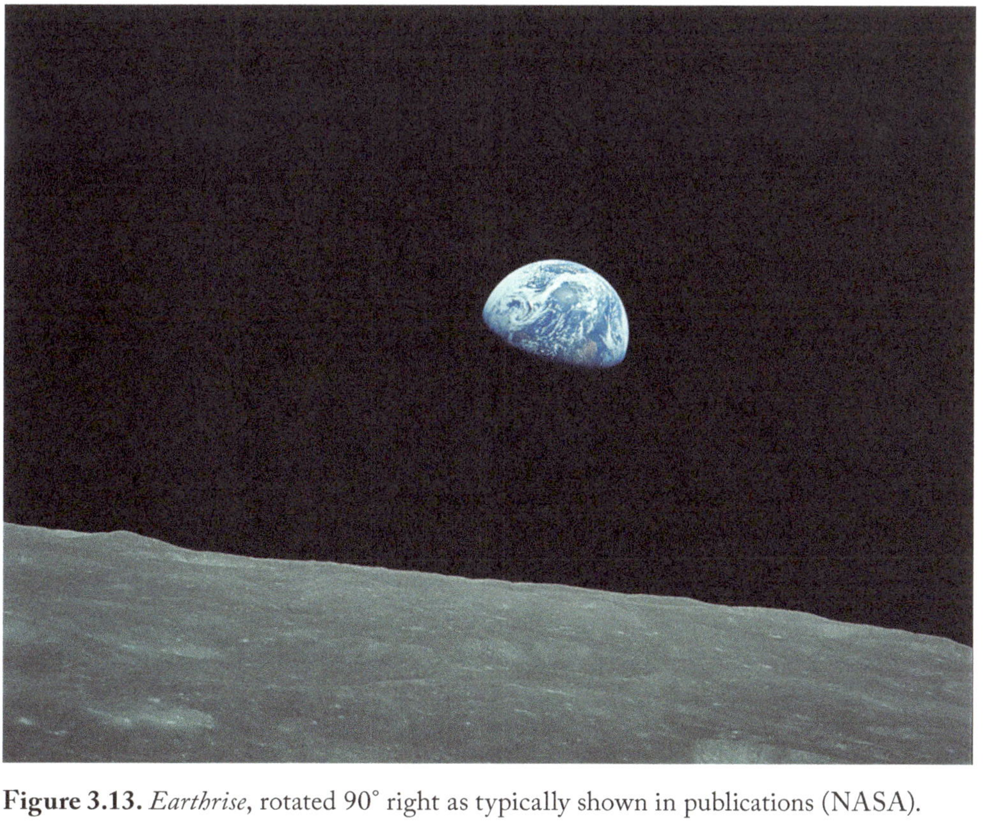

Figure 3.13. *Earthrise*, rotated 90° right as typically shown in publications (NASA).

came to understand exploration through information found in newspapers, books, magazines, and on television, media formats all heavily reliant on visual material. Production and dissemination in both of these instances, not surprisingly, had similar origins and needs. Developing a sense of meaning for these images, as it was then and now, is the interpretive step taken in part by the remainder of this section.

While thousands of photographs exist from the early U.S. space program, I selected just a few here as useful comparisons for the rich set of visual examples from other arduous and image-rich missions of discovery. As explained, a key visual counterpoint to photographs taken during the Great Surveys and *Terra Nova* and *Endurance* trips to Antarctica is the iconic image *Earthrise*.[52] The concept of the frontier, represented through alluring images of a distant horizon beyond a barren or treacherous landscape, plays a significant rhetorical role in many of the expeditions. While the visual cues are more than obvious, other elements of the image's context make for a worthwhile deeper dive into this case study. We see how persistent certain themes are throughout photographic history.

If NASA, like corporations and governments before it, cared little for the aesthetic qualities of the images, how do such images become such an important part of an American visual legacy beyond the space program? It bears noting here that the quality images from the Great Surveys, while unable to rival the picturesque qualities of painted canvasses by Moran or Alfred Bierstadt, were taken by professional photographers who carefully composed their images. The ability of professional photographers to conceive of a scene, a factor in the earlier expeditions, was likely not a factor in considering the aesthetics of photography during the Apollo program—there was no room in a tiny capsule for an additional person to take pictures. Astronaut photographers were there to prove the technology of spaceflight worked, and farther down the list of priorities, document the experience.

Nationalistic overtones present in the rhetoric of the American Great Surveys and the British Antarctic trips spill over into the visuals, though more subtly than one may see in images of astronauts on the Moon. An image from Apollo 15 of Commander David Scott saluting the U.S. flag firmly planted in the lunar surface (Figure 3.16) echoes jubilant images from western exploration (Figure 3.14) and the Antarctic (Figure 3.15). While not as explicit as the Apollo 15 image with the prominence of the U.S. flag, the other figures express satisfaction with the accomplishment. Despite possible problems or failures in their mission, these images offer a sense of human scale in relation to landscape features otherwise indiscernible to the audience.

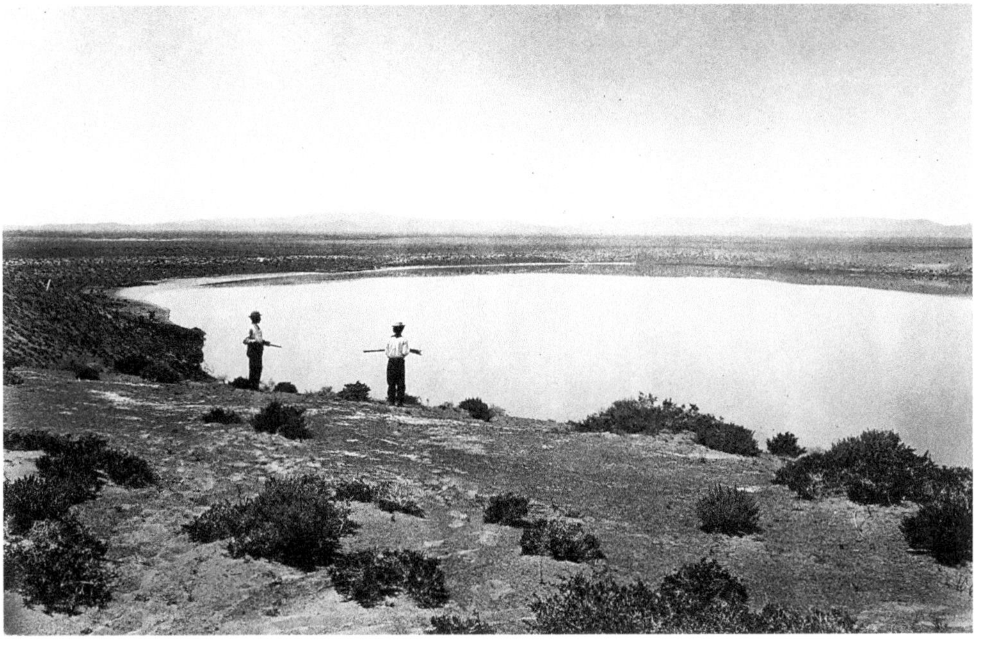

Figure 3.14. Alkali Lake, Carson Desert, Nevada, photographed by Timothy H. O'Sullivan, 1867, LC-USZ62-103091 (Prints & Photographs Division, Library of Congress).

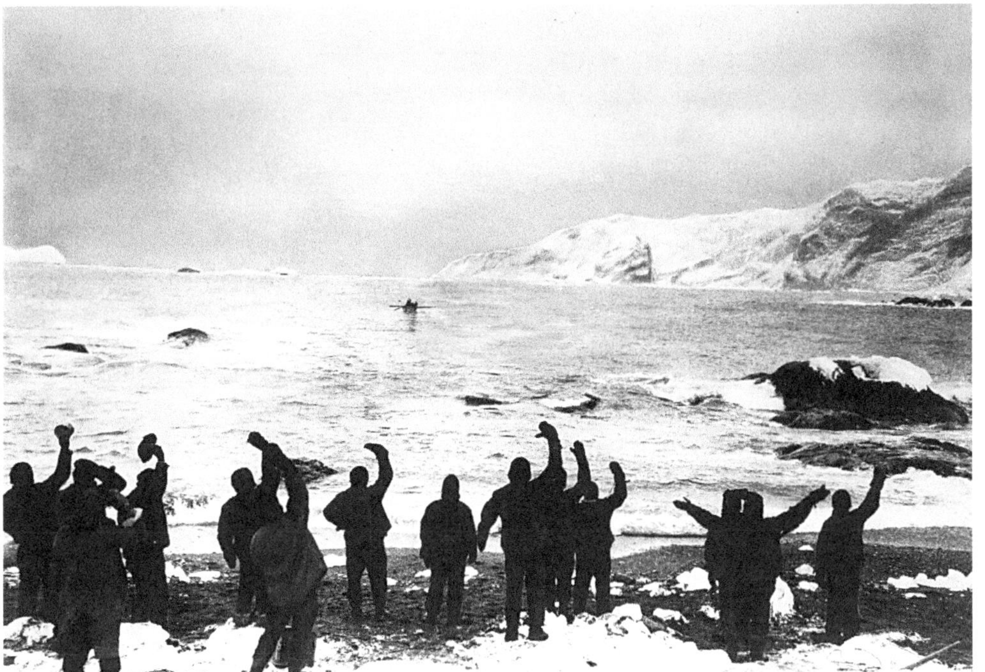

Figure 3.15. *Saved!* Frank Hurley, 1917 (Scott Polar Research Institute, University of Cambridge).

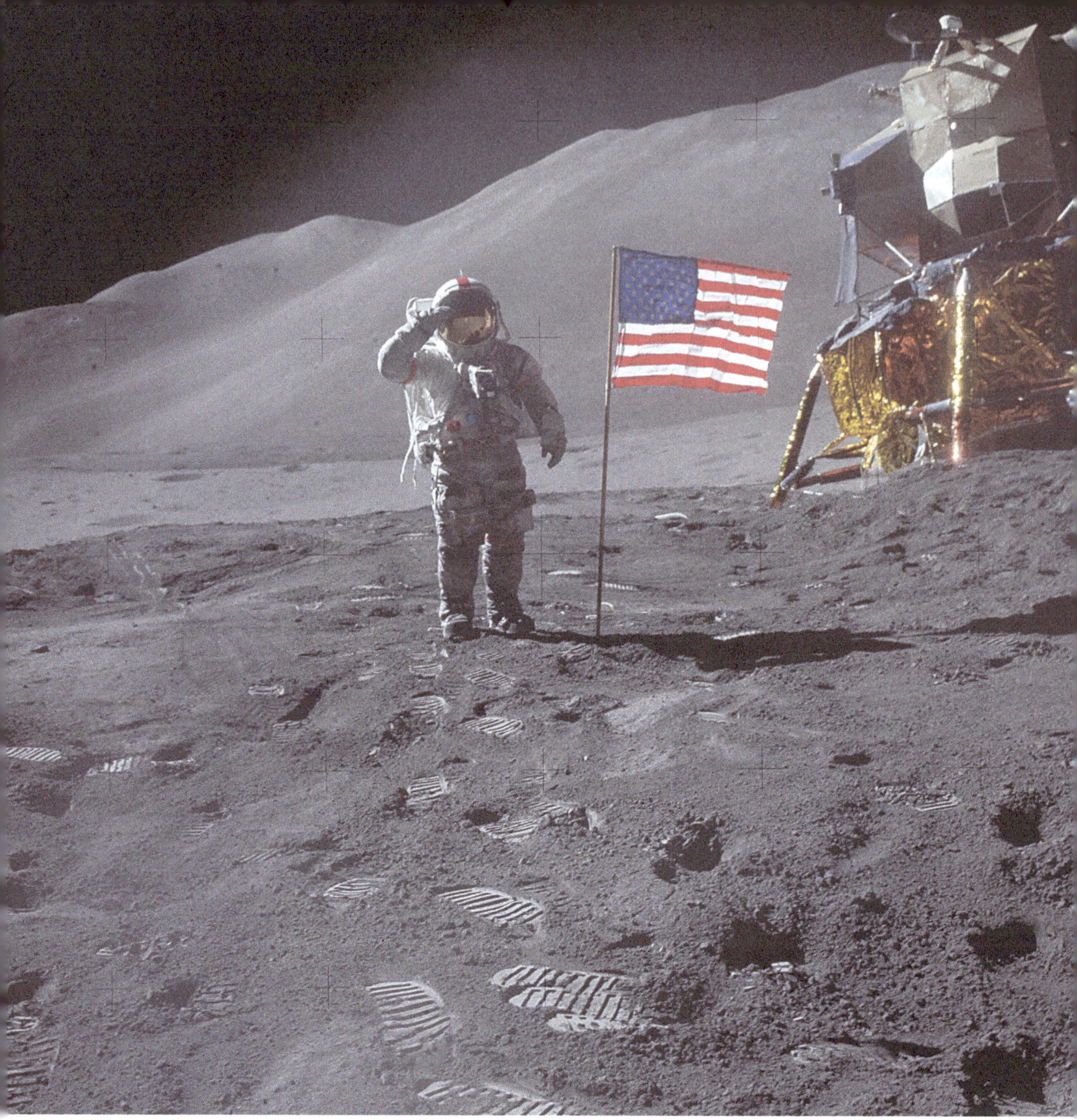

Figure 3.16. Astronaut David Scott on the lunar surface during the Apollo 15 mission, August 1, 1971, AS15-88-11863 (NASA).

Triumphing over unknown lands and people such as British empire-making provides a substantial point of comparison with the American victory in the Space Race as seen in *Earthrise*. Such imperialistic projects provided visuals such as "photographs—composed, reproduced, circulated and arranged for consumption within particular social circles in Britain—[that] reveal as much about the imaginative landscapes of imperial culture as they do about the physical spaces [pictured]."[53] Images of success serve the needs of expansion and the nature of those doing the conquering. Anne Lindbergh astutely recognized the space-African exploration connection as well in the preface to her "Earth Shine" essays. In compiling a book of essays about space and traveling in Africa, she notes that the two places are joined by their extremes, reminding the reader that "We cannot live in the wilderness and forget civilization. Nor can we live in the heat of civilization without realizing its increasing problems."[54] In photographing *Earthrise*, William Anders unknowingly reinforced the long-held idea that the Space Race was the ultimate attempt to expand using Manifest Destiny principles of decades past.[55] As a signal of victory, *Earthrise* fits into the imperialist genre of photography, as well as maintaining an association with the concept of the frontier so pervasive in American visual culture.

Setting a human in the context of harsh environments, places rather unknown to audiences, provided a sense of understanding, but belies a number of realities of the situations depicted. While the waterfall-like setting of William Henry Jackson's images of a survey companion standing on Mammoth Hot Springs in Yellowstone Park looks elegant and conveys the natural beauty of the spot, the danger of scaling the terraces and avoiding the scalding water (averaging about 80° C) put the photographer and survey member in considerable danger (Figure 3.17). Danger abounded in the Antarctic as well. Robert Scott grew concerned about Herbert Ponting's risk-taking for photographs after attacks by penguins and a near fatal encounter with orcas he attempted to photograph from an ice floe.[56] The ice presented obstacles at every turn for polar expeditions, but were ready-made photographic subjects for giving a sense of scale as well (Figure 3.18). Geological features of the lunar surface confronted Apollo astronauts. Known for its wide-ranging examination of the Taurus-Littrow lunar highlands, the Apollo 17 mission plan took astronauts Eugene Cernan and Harrison Schmitt through what some suspected was the site of recent volcanic activity (Figure 3.19). They found numerous large boulders, unusual soils, and returned more samples to Earth than any other mission. Considering the low-gravity, airless environment in which

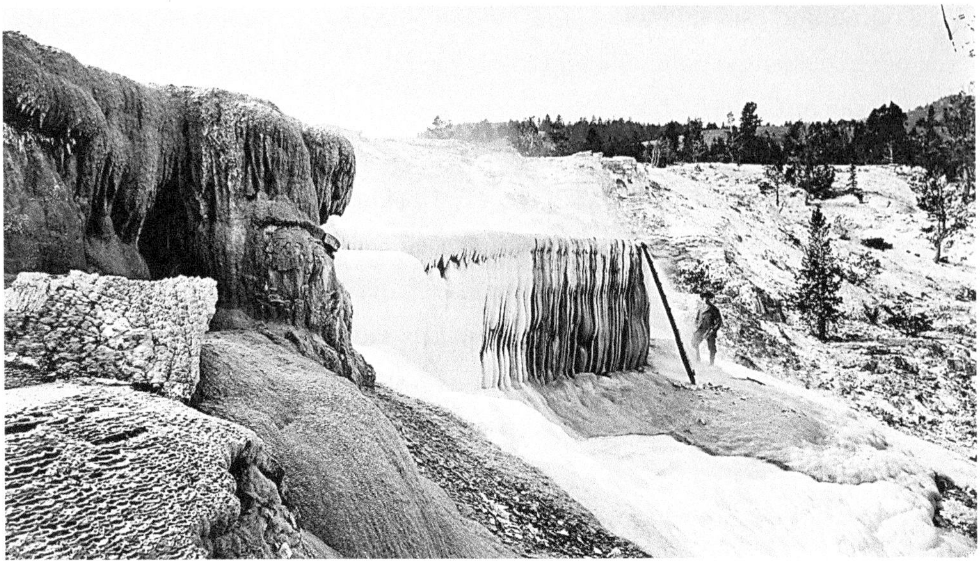

Figure 3.17. Mammoth Hot Springs, the Frozen Waterfall, Yellowstone National Park, photographed by William Henry Jackson, 1878, 57-HS-362 (National Archives).

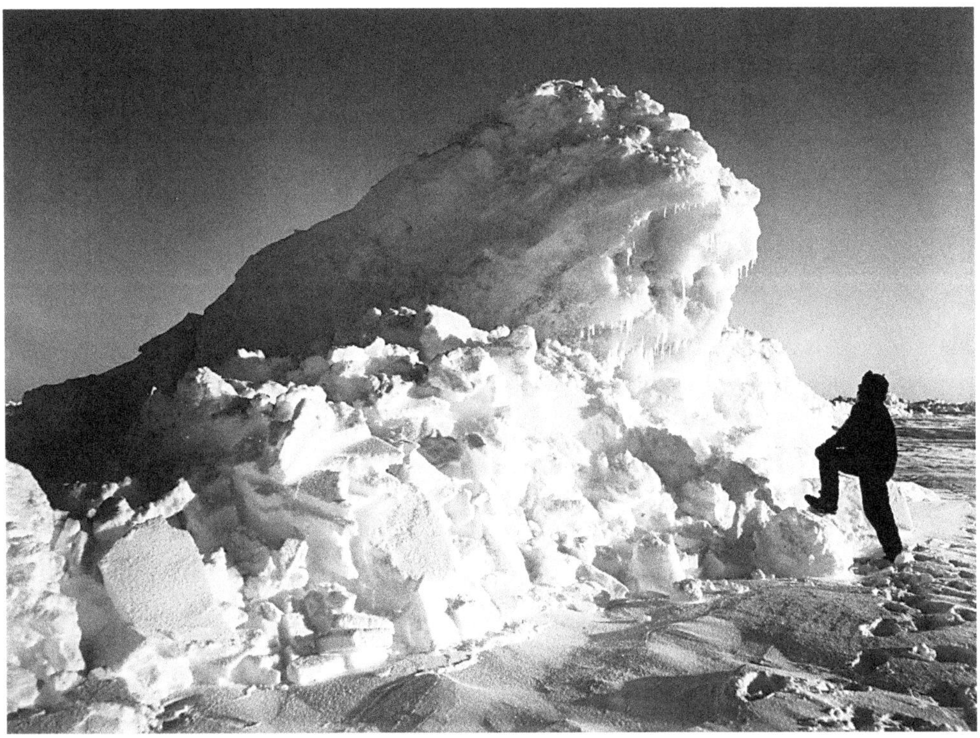

Figure 3.18. A typical pressure ridge with rafted floe, photographed by Frank Hurley, 1914–17, P66/18-40 (Scott Polar Research Institute, University of Cambridge).

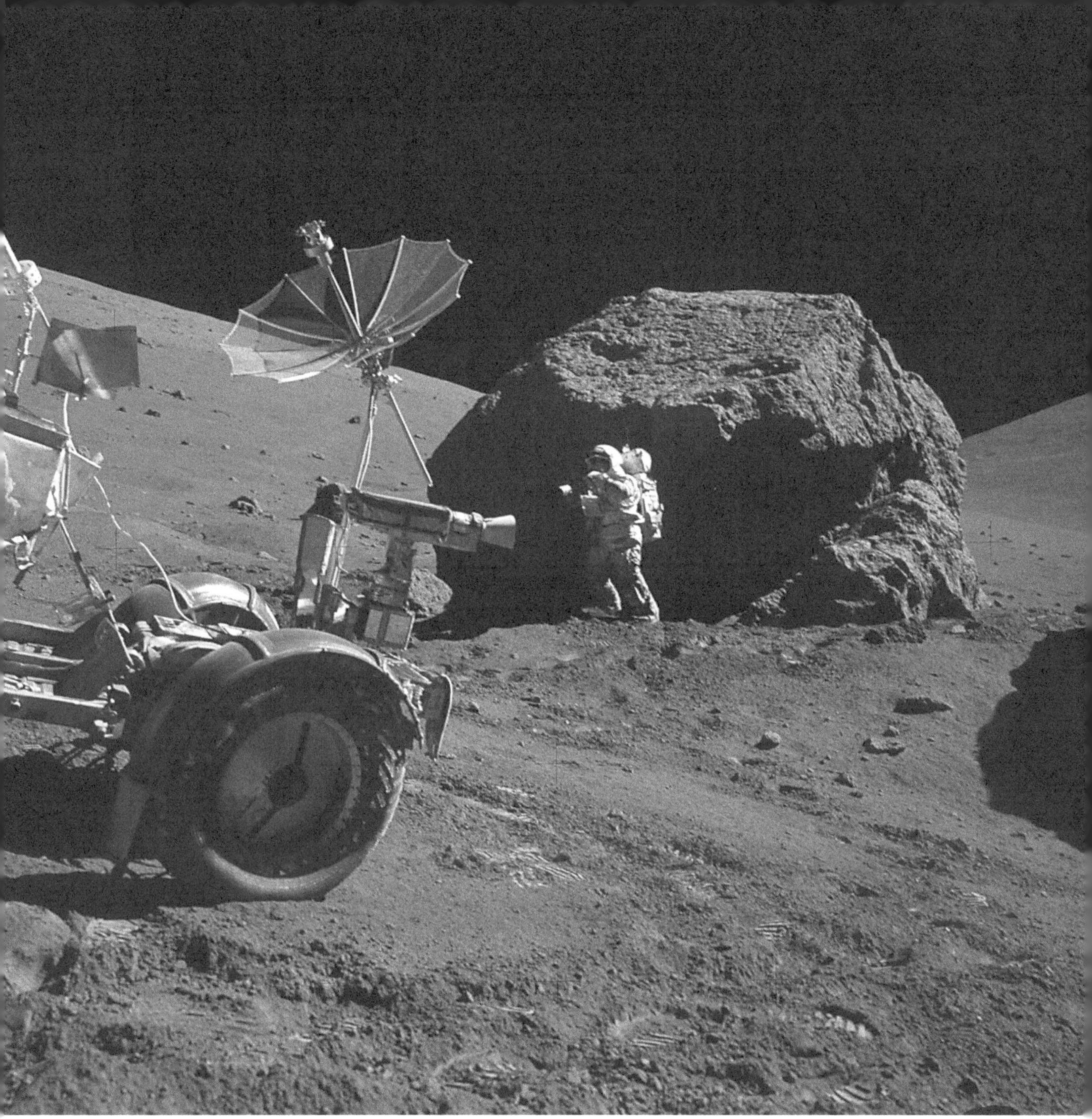

Figure 3.19. Astronaut Harrison Schmitt standing next to a boulder during the third EVA of the Apollo 17 mission, photographed by Eugene Cernan, December 13, 1972, AS-17-146-22294 (NASA).

they operated, astronauts faced greater challenges that deterred further human exploration of the Moon than those of the past, and photographs of humans in that place served to confirm those concerns.

Another common theme throughout exploration photography is what might be deemed the rearview mirror images, those that appear to consciously take a look behind the photographer to give the viewers an understanding just how the photographer and expeditions members reached these far-off locations. This strategy reinforces the concept in photography that the best image may often be behind the photographer. In cases of exploration, the mode of transportation becomes the subject of the photograph, and the photographers play with concepts of distance between the camera and the vehicle and the vehicle and the horizon. The breadth of these types of images include those focusing simply on perspective through those strictly for documentation, drawing as much attention to the camera's point of view as it does the subject of the image. Across the expeditions examined here, the importance of these images lies less in their public appeal as part of a mythical journey to the unknown and more on their place in creating a realistic visual narrative, capturing the experience for posterity.

Transportation technologies in the context of the landscape served a number of purposes for exploration photographers. Putting them in the physical context of unknown landscapes gave the tools of exploration some needed context. Reaching remote canyons via rocky landscapes and rough rivers required lugging equipment and supplies for hundreds of miles, so Hillers' depiction of transportation methods used during Powell's second trip to the Colorado River made the scale and danger of the voyage evident (Figure 3.20). Similarly, while large sailing vessels were familiar to people of the early twentieth century, their appearance in photographs by Ponting and Hurley clarified their use in transporting people and equipment to the hostile icy landscape of the Antarctic. As part of the narrative of the Weddell Sea party Shackleton led to reach the South Pole, Hurley's photographs situated the team's struggle to save their ship and supplies from the treacherous ice (Figure 3.21). Only this image and a handful of other surviving frames remained to tell the story of the party's battle in the Antarctic.

The unfamiliar appearance and function of the lunar landers resting on the dusty lunar surface did not present the same spontaneous understanding of lunar transportation as canoes and sailing ships offered audiences of the American West and Antarctic exploration photography. The sight of these fragile, human-built devices within the

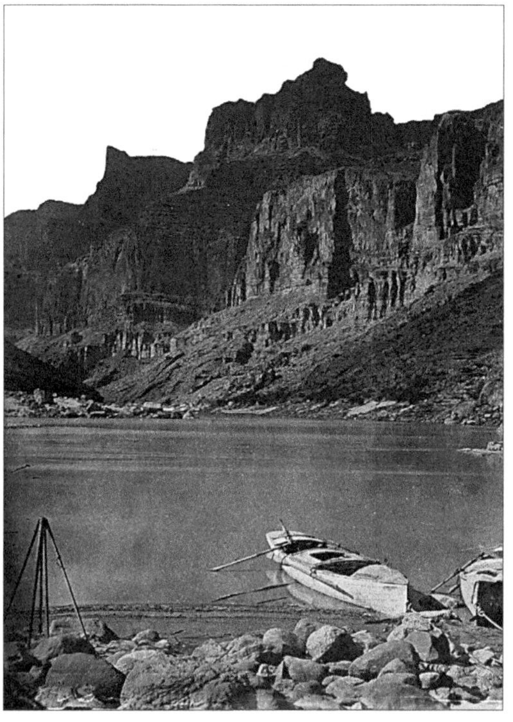

Figure 3.20. Boats near the mouth of the Little Colorado River, photographed by John K. Hillers, 1872, 57-PS-885 (National Archives).

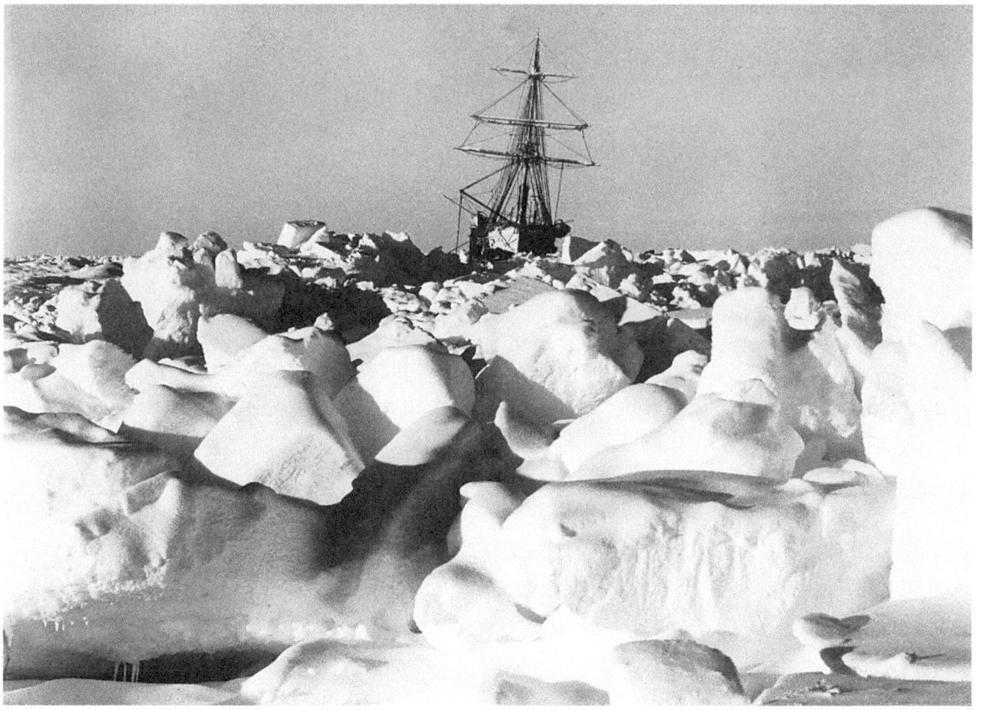

Figure 3.21. Conditions of the ice ahead of the ship, photographed by Frank Hurley, August 1915 (Scott Polar Research Institute, University of Cambridge).

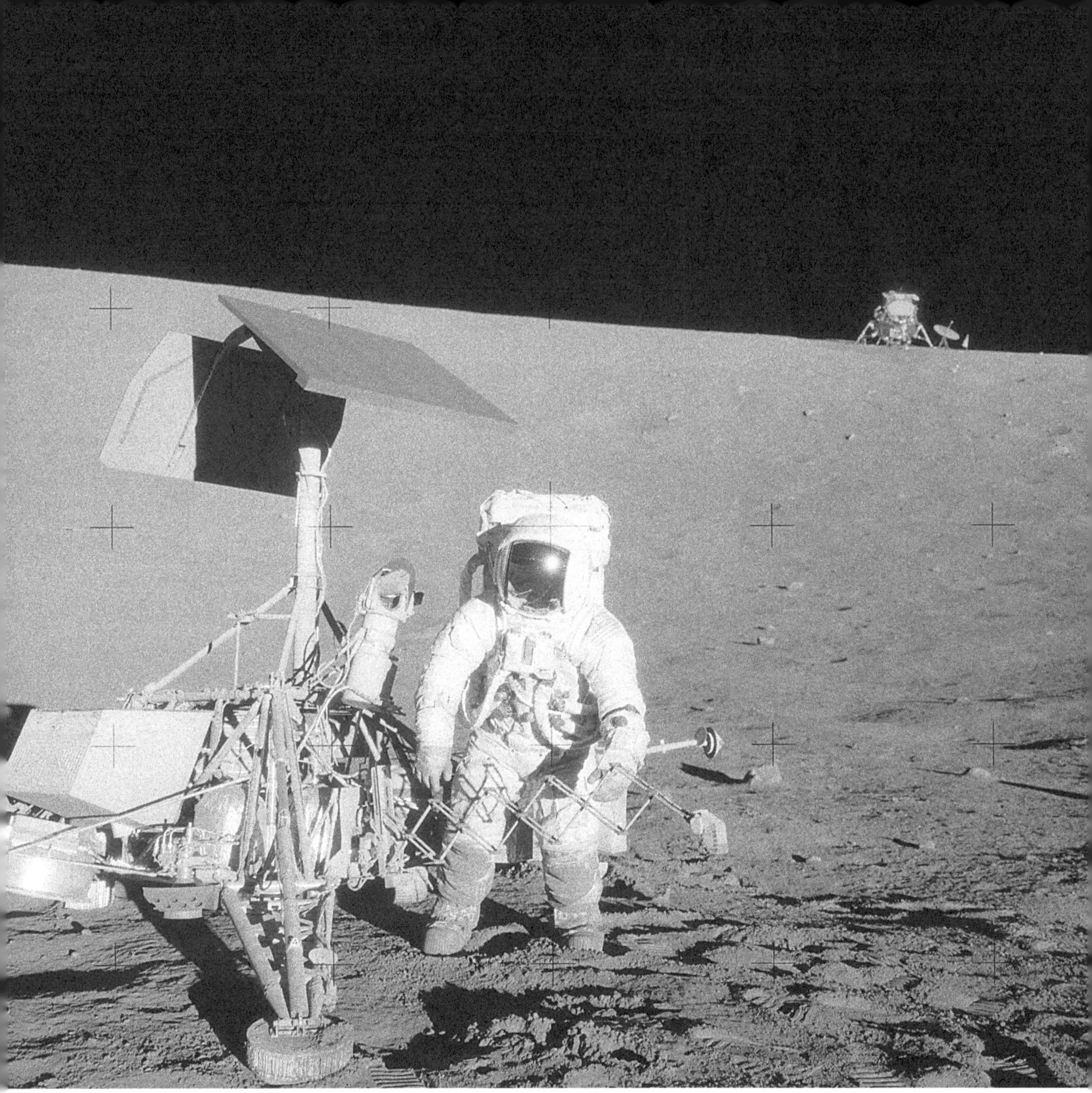

Figure 3.22. Alan Bean with the Surveyor 3 spacecraft, lunar module *Intrepid* in the distance, photographed by Pete Conrad, November 20, 1969, AS12-48-7136 (NASA).

harsh and alien landscape offered some visual clarification of the lunar landscape's danger. Astronauts in spacesuits confirmed the challenges of the Moon. In the case of Apollo 12 astronauts Pete Conrad and Alan Bean, their second moonwalk brought them to a lunar resident of Earth origin: the Surveyor 3 lunar lander, sent in 1966 to perform preparatory research for human landings. The view from Surveyor 3 back toward the lunar module *Intrepid* contextualizes the mission as one of long voyages for both the human and the robotic, displacing the viewer from traditional exploration photography only in setting, but not in perspective (Figure 3.22). The audience for space photographs, like those viewing scenes of American West and Antarctic exploration, saw a constructed scene meant to convey the distance traveled, the requirement of technology to achieve that voyage, and the harshness of the landscape in which they saw the voyagers. Perhaps even more indicative of this view of distance traveled is an image of Harrison Schmitt during Apollo 17, which contains ideas of nationalism, humanity, and technology in a single image (Figure 3.23). At no other time in the history of exploration did a photographer capture the mission's rhetoric, origins, destination, and participants as expertly as Cernan did in this photograph.

(•)

Audiences, in varying orders of priority, remain the same throughout the history of exploration photography: scientists, patrons, and the public. Reviewing the series of photographs collected during any of these three expedition periods, it is evident that the overwhelming majority of photographs collected are landscapes and nature views. Rarely are people depicted, save in the cases of special studies of Native Americans during the Great Surveys or of expedition crewmembers for the purposes of giving human scale to natural features. Even astronauts make few appearances in still photographs compared to the bulk of images captured for geological or other scientific purposes. Simply stated, people were not the most important subject of exploration photography. Considering the limited resources available to the photographers on these voyages, allocating more than a handful to showing fellow travelers would have ignored the primary purpose of using expedition funds on photography: the needs of the primary audience for those images. In each instance, patrons wanted to use the scientific or technical information available in those photographs to their benefit. Getting images to the audiences was the next step in the narrative of exploration photography.

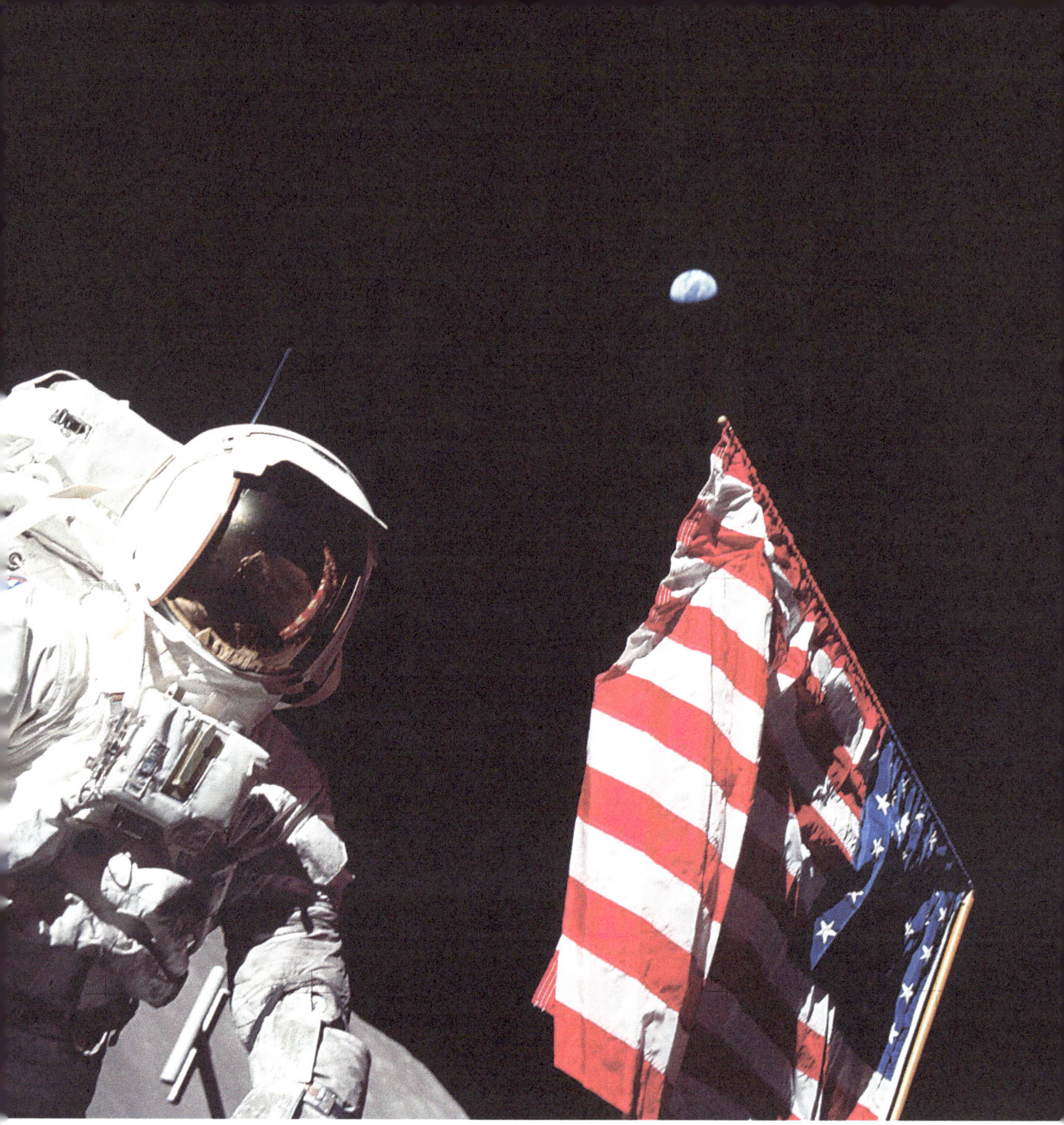

Figure 3.23. Harrison Schmitt and the U.S. flag with Earth above, photographed by Eugene Cernan, December 11, 1972, AS17-134-20384 (NASA).

The state of printmaking and reproduction technologies played a significant role in the ability of image makers to distribute images to audiences. While developing multiple prints from glass plate negatives posed no problem for late-nineteenth-century photographers, there was no means to reproduce them for mass-market publications like newspapers until the halftone process came into use during the 1880s. Audiences were limited then to those able to purchase prints (often stereographs), view them in an exhibition, or access them in a bound photographic catalog such as those distributed to Congress by geological survey leaders. A fairly selective and elite audience to be sure. While government-sponsored efforts of the Great Surveys meant that copies of the prints ended up in government repositories, copies made for other patrons and sale to the public by the photographers using their original glass plates meant a limited circulation of prints, but one that created a reputation for these visions of the West. Providing precise visual representations to a public accustomed to lithography and paintings was of little importance. Those, like Congress, who required scientific documentation to plan future development of the added territories, received the necessary reports with plenty of sepia-toned printed photos included.

Make no mistake about the intentions of these and later exploration photographers: they did not join expeditions for selfless purposes of supporting science or exploration. William Henry Jackson agreed to supply any needed prints to the government while retaining the right to sell the images he captured on the Hayden surveys. Herbert Ponting and Frank Hurley undertook extensive, for-profit lecture tours after their Antarctic adventures. Photographers on those expeditions sought personal benefits from their work. With more advanced means of reprinting images in newspapers, magazines, and books, the Antarctic photographers reached a broader public audience. The images not only became part and parcel of funding plans for the missions themselves, but also supported the long-term individual finances of Ponting and Hurley. Their status as full-fledged members of the Scott and Shackleton expeditions meant that their experiences and images contributed to a close public identification between them and those disastrous polar trips. In the cases of photographers of the American West and the Antarctic, the professional nature of the work and shrewd financial arrangements kept ownership of the photographs largely in the hands of the photographers. And as the age of heroic Earth exploration came to an end with the start of World War I, most found fame and fortune far less financially beneficial than they expected.[57]

Astronaut photographers had no such commercial angle to play or even any involvement in image distribution. As a government endeavor, the space program's images, save in instances of being withheld for national security reasons, became part of the public domain and usable by anyone.[58] The expansion of mass media and a hunger for immediate visual representations that accompanied the rise of television gave NASA good reason to circulate still images quickly and broadly. Space-related television broadcasts were fleeting moments in this period because homes lacked recording technologies. Newspapers and magazines, however, became family keepsakes, leafed through multiple times and tucked away in a cedar chest as a memento of the experience. NASA found a willing consumer for the photographs in a culture adapted to absorbing visual images with few questions. In essence, NASA sold itself to the public not through spectacular feats with hardware alone, but the ideas the iconography of space exploration could convey "an image of national purpose that equated technological preeminence with military, ideological, and cultural supremacy."[59] Astronaut photographs, in short, were the primary vehicle for the literal and rhetorical image of the space program. But to some, the ideas they conveyed held more long-term value than the instant gratification offered by amazing views of human-made technologies sent outside Earth's atmosphere.

The success of rhetorical images to accomplish a mission meant gathering an impression of actions taken by viewers. Depending on the mission goals, audience use of images included seeking funding for additional exploration, speaking to others about the images, or coming to believe in the ideology conveyed by the narrative of the images. In the instances examined here, explorers meant for their project photography to reinforce ideas established from the expedition's outset: finding natural resources (for public and private purposes), scientific documentation, increasing national prestige, or the abstract concept of the inevitable need for humankind to explore and conquer. Responses to images by audiences, such as scientists, bureaucrats, industrial leaders, other nations, and the public, increased ideological and physical attention paid toward those geographic regions.

Images alone did not drive migration to the American West, but they certainly contributed to its myth and that of the United States as a land of opportunity. With the completion of a transcontinental railroad line in 1869, the era of easier exploration, economic development, and travel beginning at stops along that line became reality. Companies looking to exploit the land could move supplies and employees there with hopes of a profit. Individuals moved there on the promise of rich soil for farming and pasturing animals. These visions, aided by the photographs, textual descriptions, and artwork generated by the Great Surveys, drove the idea of the frontier as open and inviting while excluding the harsh realities of the climate, landscape, and any existing occupants.[60] Here is where the traditional mythology about the West breaks down. No amount of careful photo editing into a narrative series accompanied by descriptions from expedition members could remove the reality of native peoples from life in the West.

Trips to Antarctica and the Moon had no preexisting populations to threaten, so developing a rhetoric relating to national prestige, competition, and scientific discovery came easy. And while accomplishing notable firsts such as reaching the South Pole or the Moon may appeal to those interested in the geopolitical stature of their country, learning about the geological and scientific character of these desolate places provided only short-term gains for the nations involved in the projects. Not only did Roald Amundsen, a Norwegian, reach the South Pole before Scott or Shackleton and secured photographic proof, but any prestige to be gained in the western world by those British failures dissipated with the start of World War I. In a strangely similar way, enthusiasm for space exploration in the United States, built on Cold War competition and the popularity of space science fiction over decades, waned when Americans saw the devastation of the Vietnam War on their television screens every night. Once won, the Moon race held little lasting interest for public audiences when the complexities of fantastical human voyages to a barren Moon became clear and were set against the reality of difficult economic times and warfare. Scientists and engineers, however, continue to study both regions to this day, learning information that encourages further exploration.

The role of images in satisfying the immediate needs of the public was accomplished by these expedition projects. Photographs of the new places showed a boundless landscape, but one conquered by heroic patriots seeking knowledge and resources. The real audiences of the photographs, those hoping to find additional reasons to support exploration and development, needed evidence and the testimony of those who took the photos for study. Without digital scanning and file sharing via the Internet, the circulation exploration photos happened slowly. Having accessible images meant that users could interpret visual documents as accurate representations of unimaginable places made real through the development of exploration rhetoric, financially supported by governments and investors, and the employment of capable technicians and technologies. Utilizing a reliable means of production permitted photography in challenging environments, guided by strategic engagement of the audience and visual planning developed by expedition leaders. While not all memorable, these images served their purpose as documentation of exploration, showing the human hand in creating new landscapes on this world and others.

Like masterful paintings that preceded it, photography contributed to a sense of national identity formulated by those venturing to the American West from eastern cities and towns. The usefulness of the landscape image lay in what Angela Miller calls "its multivalence of meaning—was well adapted to serve a diffuse nationalism without provoking more direct allegiances to place or section, to social class, or to urban polity."[61] This inherent element of visual culture, multiple, unpredictable meanings, makes any interpretation highly problematic. Attempts to then link the landscape of the West to a sense of American-ness was fraught with challenges as the landscape itself was a contested space.[62] Similar problems exist when analyzing other exploration photographs, but one can tease out messages within distinct types of photographs and their likely meaning for photographers, expedition leaders, and a general audience based on image content and context.

Beyond issues of process and distribution to audiences, the looming issue for analysis of exploration photography is determining what meaning they hold for audiences: the means of consumption and eventual placement of those images in our collective memory. Prints from either glass or acetate negatives or reprints in publications gave photographs a lasting accessibility far beyond that of paintings. The availability of photographs gave audiences repeated opportunities to interpret and find their own meanings in the photographs, however much the images were guided by captioning or other textual references. Artists, scientists, engineers, government administrators, and casual viewers had the opportunity to find within those photographs what they wanted to find based on their own needs. Apollo images, however, far surpass all prior exploration efforts thanks to advances in technology that allow repeated and often unexpected uses in the decades since we last went to the Moon.

CHAPTER 4

The Afterlife of Astronaut Photography

> *"[W]e've only seen ourselves through the paintings of artists, words of poets or through the minds of philosophers. Now we've been out there, we can see ourselves."*[1]
>
> *—Eugene Cernan, Commander, Apollo 17 Commander*

As part of the opening minutes of the Oscar-winning documentary *An Inconvenient Truth* (2006), former Vice President Al Gore presented his audience with two iconic space images: Apollo 8's *Earthrise* and Apollo 17's full Earth, often called *Blue Marble*. Although his description of *Earthrise* is inaccurate and overblown (he calls it the first picture of Earth from space that anyone ever saw), Gore situates it appropriately as a turning point in American public memory and move toward intense environmentalist activity. His brief narrative about the full Earth image provided precise details but failed to identify the image as a literal banner of Earth Day for over fifteen years prior to his film's release. In a film intended to convey the consequences of inaction to combat climate change, then forty-year-old photographs still held sway as symbolic of Earth as a fragile and beautiful place in contrast to a stark lunar landscape or the blackness of space. Gore's presentation, an environmentalist interpretation of images captured by astronauts as part of their assigned duties, reinvigorated interest in NASA photography just as electronic resources for viewing images became available. The writers behind the film were hardly the first to employ these images as historical markers of an age, yet only a handful of scholars have considered their impact and resonance.

NASA as an organization may not have spent much time considering the potential effect of astronaut photography on the agency's legacy, but public impressions of images, the collective memories of early human spaceflight, reverberate through science and popular culture to this day. Until high-resolution images from the DSCOVR satellite returned a full Earth image in June 2015 from Lagrange point 1 (L1), no other single frame of film existed to show the earth as seen in the full Earth image from Apollo 17. For forty-three years, a single image represented our conceptions of Earth at reasonably close range and meant that even for those with no living memory of the early 1970s, the so-called *Blue Marble* was Earth for everyone. Forming at the juncture of mass culture, personal narratives, and photography, this astronaut-captured view formed what Alison Landsberg calls a *prosthetic memory*, facilitating deeply political and ongoing connections to something viewed but not experienced. All those years of circulation of the Apollo 17 image means that it may take some time for a new *Blue Marble* to surpass its predecessor in terms of a prominent place in public memory.

To achieve a lasting place, NASA images of the 1960s required professional production and adequate distribution to meet audience expectations. Exploration of the past included professional photographers, and without room inside a spacecraft for one, astronauts served as their proxies. The images identified most easily as symbolic of NASA's successes became so in part because of their relationship to an existing visual lexicon of exploration. Space might have been a new place for humans, but our expectations for what we would see was shaped by the visual rhetoric of exploration. The elevation of these images to iconic status happened because of the reproduction of the images over decades and encounters with them that created sometimes deeply personal responses.[2] With images in hand, audiences could use and reuse them in countless ways to celebrate and imprint upon our collective memories the association of NASA and its astronauts with ideas of scientific and technological dominance.

BECOMING ICONS

For decades leading up to the final mission of the Apollo program in late 1972, moviemakers, artists, and writers speculated on how such voyages to space might shift human perspective. Would we establish Moon colonies for later trips to Mars and beyond? How would this intimate knowledge and experience with space travel affect our conception of our place in the universe and inspire travel to other planets? Or, as some speculated, would humankind turn attention back towards the Earth itself with the ability of humans and satellites to image the planet in parts or as a whole orb? Or as some counterculture luminaries asked, "Why haven't we seen a photograph of the whole Earth yet?" (Figure 4.1). One outspoken writer of the counterculture movement of the 1960s, Stewart Brand, drew the attention of many around the world toward the latter perspective with his environmentally minded campaign to encourage the publication of a whole Earth image. Brand's work, in addition to that of the growing environmentalist agenda spurred on by works such as Rachel Carson's *Silent Spring*, brought about a dramatic shift in the public perception of astronaut photography during the Apollo program.

Coincidentally to the growth of the military-industrial complex of the 1950s came a growing concern for the welfare of the planet. Engineers and scientists manipulated materials into new chemicals for industrial, commercial, and military uses, causing many to question the potential danger such things posed to Earth and its inhabitants. Carson, a marine biologist, voiced in print the conservationist cause. That found sympathy in the counterculture movement, where many were concerned with a perceived aggressive commercialization that threatened their ideas of a peaceful and environmentally friendly society.[3] Brand was one of those who took up the flag of Carson's concerns, inspiring budding environmentalists to think about how NASA could play a role in shaping the future of their movement.

As Brand himself explained, he was in the middle of an LSD hallucination when the San Francisco skyline inspired a thought about the curvature of the Earth: why was there no published photograph of a whole Earth?[4] Satellites and other spacecraft had yet to be equipped with a camera and the necessary technology to capture a photograph of more than partial Earth at the time of Brand's "trip" in 1966. What he imagined, and used to energize his campaign, was the potential power of an image depicting the entire Earth against the black backdrop of space. For the surging environmentalist movement and a counterculture concerned with peace around the world, a depiction of the globe and characterization as a fragile place needing protection served as a rallying cry.

Brand's cause became one of invoking what he called the "great American resource of paranoia" by turning the demand for the photograph into the question: "Why haven't we seen a photograph of the whole Earth yet?" Plastered on the ubiquitous buttons and posters common to counterculture life, Brand literally sold his message to the American public of his own initiative. He sold them on college campuses, first at the University of California at Berkeley, and then others around California and the country. He and his friends sent them to scientists and anyone he felt had influence with NASA or the Soviets to turn cameras back towards Earth during robotic or human missions.[5] Having heard his plea for a whole Earth image, NASA, not unfamiliar with the concept of taking such a photograph, used its geostationary weather and communications satellite ATS-3 to give the world the most complete photograph yet of the Earth in late 1967. Once released, Brand had the perfect image for the cover of the very first issue of *The Whole Earth Catalog* (Figure 4.2). Building off the successful community response to the buttons, Brand and his cohort published their first issue in the fall of 1968, a guide and retail publication for products deemed useful within the do-it-yourself counterculture. The catalog's success proved Brand's campaign had real influence within the counterculture movement and made government officials take notice. Later catalogs featured new photographs of Earth from space as astronauts took them during the Apollo program, though the publication never featured the most famous of those, the *Whole Earth* image from Apollo 17. Brand's point about the value of seeing Earth from space became clear in the countless reuses of astronaut-captured Earth images in the decades since.

From the start, Brand and his cohort were critical of the lack of images of a whole Earth, but the *Whole Earth Catalog* revealed strong support for NASA and particularly astronaut photography. In the spring 1969 edition, one page contains

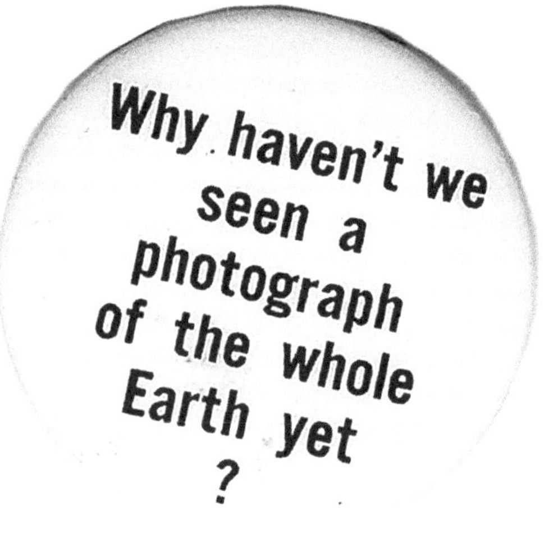

Figure 4.1. Stewart Brand's slogan for the *Whole Earth* image campaign was featured on buttons and posters like this from the collection of the National Museum of American History (Stewart Brand).

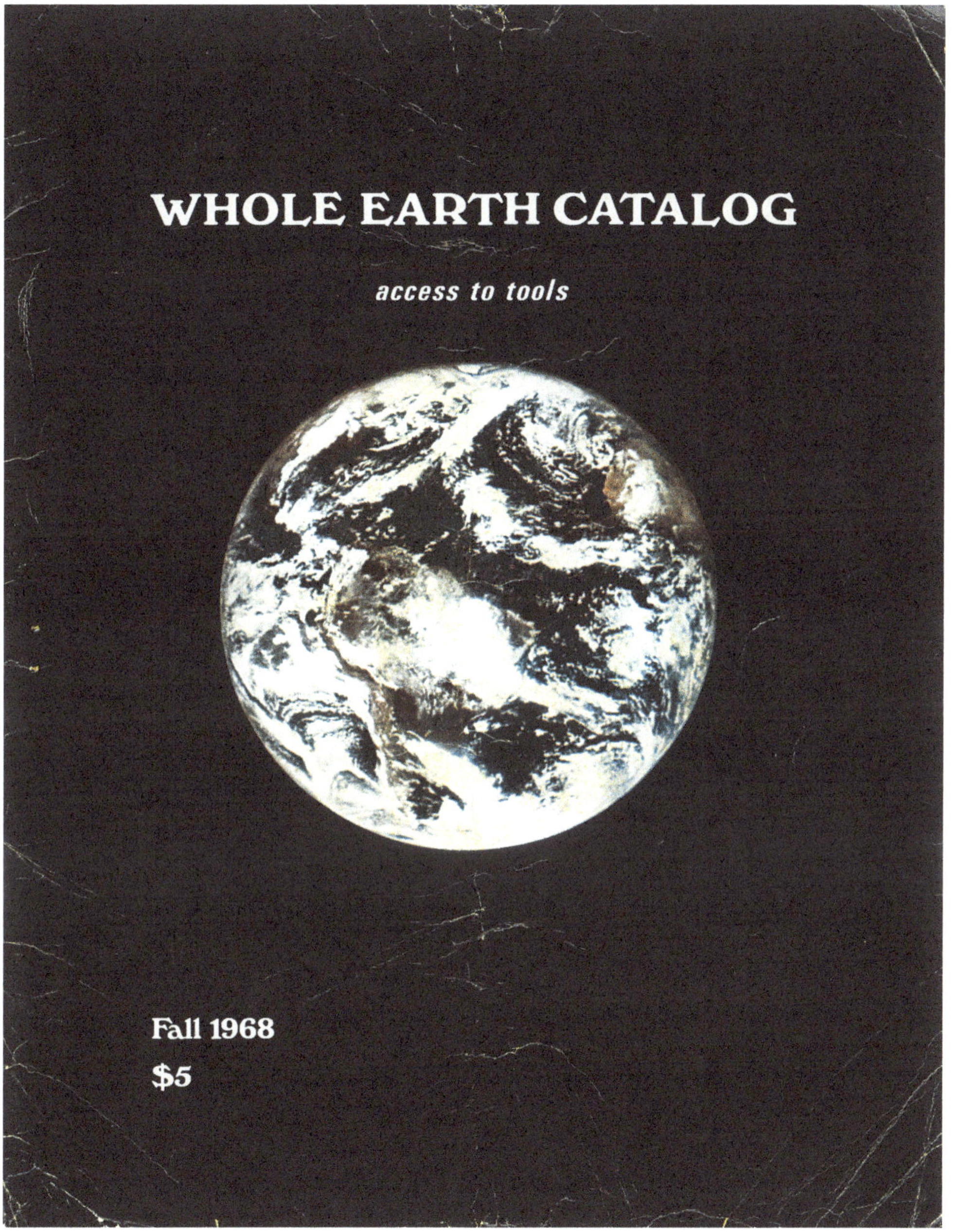

Figure 4.2. *Whole Earth Catalog* front cover, fall 1968 issue (Stewart Brand/Kentucky Historical Society).

a small feature on the Apollo 8 mission as well as a list and purchase information for recent NASA publications featuring such photographs.[6] Placed within a section about whole systems, and in the context of a publication intended to offer tools to improve understanding of and working with the environment, a feature about NASA publications and projects brought photographic accomplishments to an audience already highly receptive to the view of Earth presented by the astronauts. By speaking to the community referred to as the New Communalists, the *Catalog*'s editors made NASA astronaut Earth photographs far more accessible to an interested community.[7] Stewart Brand and those associated with the *Catalog* found that technology in the form of NASA human spaceflight and astronaut photography offered a means to promote their dream of a more inclusive, united globe. In the end, NASA needed Brand, and Brand needed NASA.

The reach of still images and the ease of their reproduction extended the possibilities for dissemination and impact across the globe, far beyond art or moving images. Photographs distributed by NASA appeared in newspapers around the nation and the world, and the United States Information Agency (USIA) used them as tools for conveying the U.S. space program's benefits to all people.[8] In an interview played on the CBS network the day after Apollo 11 landed on the Moon, July 21, 1969, former president Lyndon B. Johnson mentioned that one of his last acts as president was to pen a letter to the leaders of every nation on Earth. Enclosed was a reproduction of the iconic Apollo 8 *Earthrise* photograph.[9] He characterized the response from world leaders as amazing, noting that Ho Chi Minh's thanks were particularly meaningful to him since the ongoing conflict in Vietnam and his inability to solve it peacefully contributed to his decision not to seek reelection in 1968. By selecting an astronaut's photograph to represent his outgoing message to the world, Johnson sought to link his scarred legacy with the generally positive perception of the American space program. Johnson's actions revealed not only his pride in the accomplishment, but also the enduring power of photographs as a means of global communication.[10]

This wordless power gave some astronaut photographs, like other images widely distributed via news media, a special status: "Iconic photographs provide an accessible and centrally positioned set of images for exploring how political action (or inaction) can be constituted and controlled through visual media."[11] The iconic subset of astronaut photographs gave NASA a source of visual power, but the bulk of the images offered longevity and a legacy to the agency and its professional audiences. The example of Johnson's use of *Earthrise* as a public relations tool is one clue of the importance of images to the space program as a whole, but also the influence they could have on global opinion and national prestige.

Personal impressions of seeing Earth from space, often related in debriefings and interviews by Apollo astronauts on the nine missions to the Moon, set the stage for the dissemination and use of photography from those missions. For these otherworldly experiences, the images were crucial to creating an understanding back on Earth about the experience and value of spaceflight. While characterized accurately as documents of scientific and technical work, what set certain images apart was the discovery of the unexpected in single frames of film, things that evoked emotion in the familiar or truly sublime.[12] Stories repeated from the time of the missions and in popular histories in the decades that followed added a mystique to still images, giving them a tinge of the immortal quality photographer Richard Underwood said astronauts needed to aim for when taking pictures. How NASA and the astronauts discussed photographs played an important role in how audiences interpreted the wealth of still images captured during the early human spaceflight program.

Live television broadcasts from the orbit of the Moon during Apollo 8, almost a week before public release of the photograph itself, gave television viewers their very first human perspective on Earth from the vicinity of the moon. The pictures made Earth look like not much more than a big ball of light[13] (Figure 4.3). Although space enthusiasts watched in rapt attention, lasting memories came from the convergence of seeing this on television and printed in mass media. While NASA knew what to expect visually as the Earth emerged from behind the Moon, few could foresee the deep significance the image assumed in the decades after.

So how does having two representations—one fleeting on television and one easily reproduced in printed and collectible materials—influence the dissemination and public understanding of the color *Earthrise* photograph? One possibility is that NASA focused so much on the television broadcasts as tools for public relations that it failed to come to terms with the potential influence and emotional value of still photography from Apollo 8 or any human mission. Having access to a new technical capability such as live television, NASA certainly wanted to latch onto the public popularity of instantaneous moving images of newsworthy events. More and more families at the time were getting their news about events like the war in Vietnam through television news broadcasts instead of printed media. Putting

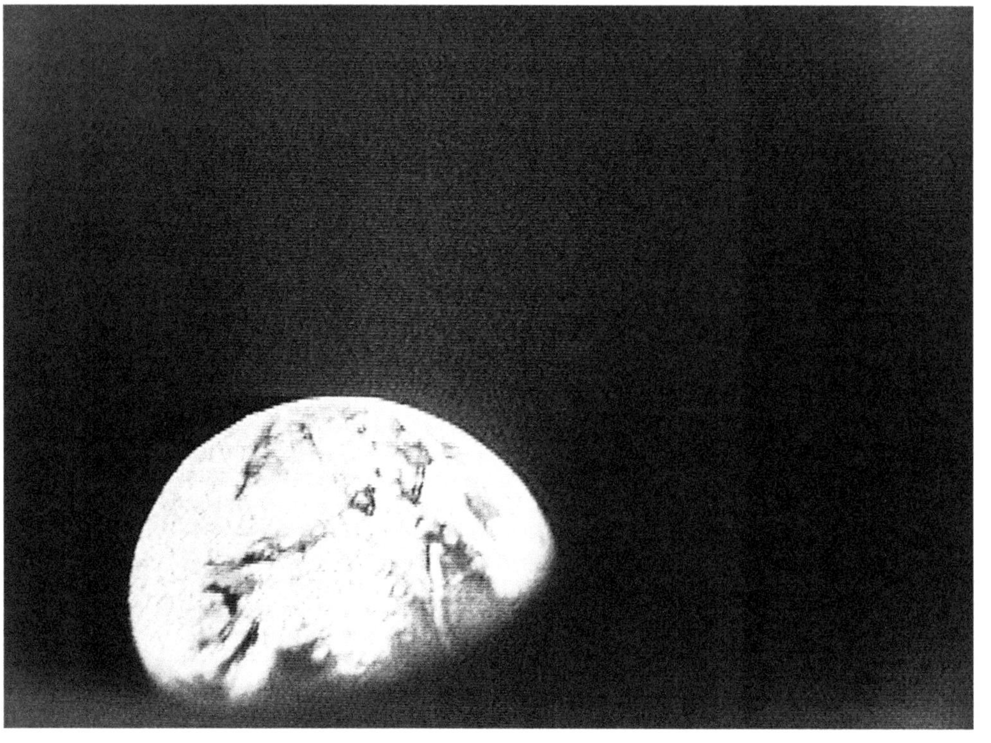

Figure 4.3. Television broadcast view of Earth from space, Apollo 8. Taken from *Debrief: Apollo 8* (NASA).

Apollo missions on display, especially this first one around the Moon, in the homes of millions around the world appealed to NASA as a means of garnering political and public support.

Examining the agency's public affairs and administrator's office records in detail, a task for future research, may reveal a different perspective on prioritizing different image types, perhaps illuminating a reason why the color *Earthrise* received less immediate attention. The story of dissemination, the last moments NASA controlled the images, is critical to analyzing their transition from only documents of a mission to visuals implanted permanently in public memory. Some sources, however, do reveal how NASA's photographic program and public affairs office disseminated still photographs. As described in formal reports and oral histories, the Photographic Technology Lab in Building 8 at the Manned Spacecraft Center (MSC) took great care in processing negatives, preferring hours of careful development over using advanced machines to speed up the creation of masters for release to the press. Underwood described the process of releasing images as rather informal, with senior managers and public affairs officers examining photographs as they came off the machines, often caught up in their sublimity. Framing these photographs with textual descriptions also shaped public understanding, though "The captions supplied by the photographic service were, however, informative rather than poetic."[14] This descriptive element, the reading of the image supplied by experts, avoided any judgment or emotive expression and allowed viewers to find their own meaning and emotion within the image.

In the days, weeks, and months following NASA missions, images spread across the pages of newspapers, magazines, and other media, sometimes alone but usually set visually within the wider context of contemporary events. As news stories and magazine features about early spaceflight missions faded, the rhetoric about spaceflight continued, amplified using photographs. Those moments became touchstones in a time of tumult, as shining moments of brilliance. A community of space historians, scientists, and enthusiasts continue to use astronaut photography even today. An equally fervent community of Moon hoax supporters uses the photography to support their own cause.[15] The apparent value of such images continues to inspire new generations of astronauts, environmentalists, and politicians who seek to revive the emotions evoked by what some see as humankind's greatest technological triumph. While lunar geologists continue to use Apollo images and lunar samples for research purposes, the life of these publicly available images continued longer than expected.

IMMEDIATE REACTIONS

As exciting and dangerous as human spaceflight appeared to those watching around the world, the U.S. and Soviet programs were far from the only thing attracting attention at the time. By the time of the final landing on the Moon in late 1972, animosity over the war in Vietnam had grown to a fever pitch with protests at university campuses and in cities around the country. Rioting throughout the 1960s over poverty and racial inequality only subsided slowly after the deaths of Dr. Martin Luther King Jr. and Senator Robert Kennedy. Ongoing troubles in the Middle East, a vibrant counterculture movement, and economic crises spent weeks and months on the front pages of major newspapers. The reality of the day-to-day for Americans rarely if ever included pondering the benefits of spaceflight, although momentary excitement over major achievements never escaped notice.[16]

Astronaut photography mattered most in the daily lives of the audiences who expected to benefit from it: NASA scientists, engineers, and management responsible for boosting the visibility of the agency's accomplishments and with an eye toward future budget allocations. Principal investigators for photographic experiments during Gemini and Apollo examined copies of returned films and published results in professional and popular journals. Frequently, analysis of astronaut photography appeared in official NASA publications, the same ones later advertised in the first issues of the *Whole Earth Catalog*. Engineers examined photographs in the immediate aftermath of missions for adjustments to equipment and procedures for future missions. Administrators presented them as proof of NASA's triumphs and mission successes. Professional audiences, the ones responsible for preplanning astronaut photography by contributing to photographic plans, used the images as evidence and influenced others within their respective fields. None of them had the ability to project or shape responses of the nonprofessional audience, those who used astronaut photographs in a multitude of media and as visual representations of their own messages about what astronauts saw.

Giving the images public meaning came in the process of their dissemination to a popular audience. Not only did those in the U.S. see film and video from missions on national and local news broadcasts, but newspapers and magazines regularly ran stories about NASA featuring astronaut photography. Thanks to the ease of spreading images via print and television reproduction, people could virtually step into the shoes of astronaut photographers no matter their location in the world. From the early days of Project Mercury, NASA worked with the United States Information Agency (USIA) to ensure distribution of images through global outreach, direct connections with the foreign press, movie and radio program production, and tours of spacecraft and astronauts.[17] The appearance of astronaut images in everything from art to lectures about the environment serves as evidence of the legacy of astronauts as our surrogates in space. Its continued use through the present day illustrates the veracity of Richard Underwood's reminder to astronauts as they trained in photography: they would achieve a kind of immortality through their images.

Gauging immediate reactions to astronaut photography by audiences requires merging information gleaned from NASA archival documents, scientific and technical reports, and some of the most widely circulated contemporary newspapers and magazines. For audiences with access to those sources after missions—days to months later—visual material played a transformative role in terms of research, technology, and personal viewpoints. For many, astronaut pictures seen in those moments solidified a public view of spacefarers, NASA's human spaceflight program, and their accomplishments. Positive public memories of those achievements took shape thanks to the visual and textual rhetoric of NASA, journalists, and others writing for public audiences. Those memories are reinforced and endure even today thanks in large part to the impressions made upon seeing the still images—and not all people did see the images—brought back from space by astronauts.

Professionals

After processing, some of the first people to review astronaut photography were the professional community of scientists and engineers working on the next steps in human spaceflight mission planning. Information contained in those images could potentially alter operations, safety, or planned experiments, so they were taken seriously. Mission planning work required capturing high-quality photographs, so it received serious attention as a part of astronaut training for data collection. The intelligence community also sought out astronaut photography in order to maintain some level of control over any high-resolution images that could pose a national security threat or provide them with useful intelligence about other countries. Inside and outside government agencies, professionals sought to use images captured by astronauts to enhance their understanding of the Earth, the Moon, and the impact of human activity on both bodies. Despite the importance of such images to researchers, little of their work made an impact on the general population.

For scientists at research universities, laboratories, and government facilities, astronaut images were an invaluable source of information for their geological, meteorological, and astronomical studies of the Earth and the Moon. Those photographs became illustrations for countless articles, presentations, and books published for decades afterward, providing inspiration for additional research and investigation with more powerful research tools in later years such as satellites and telescopes. In the foreword accompanying the publication "Manned Spaceflight Experiment Symposium, Gemini Missions III and IV," the editor states that this was the first symposium in a series intended to cover the results of mission experiments.[18] The immediate audience and participants were to be the scientific community, but as George Mueller stated in his opening remarks at the conference, NASA hoped this series would also fulfill the requirement in the National Aeronautics and Space Act of 1958 to disseminate information about space research to as wide an audience as possible. Presentations on photographic experiments used multiple images as visual evidence, including indicating which of those photographs could serve as starting points for further research. Dr. Paul Lowman cited an image from Gemini IV in his report and noted how features seen in the North African country of Chad were proof of the utility of space photography of remote locations, as well as how images could reveal previously unknown geological features.[19] Kenneth Nagler and Stanley Soules made similar forward-looking statements in their report at the conference on the S-6 Synoptic Weather Photography experiment. For meteorologists, the photographs may not have provided any dramatic evidence of unknown phenomena, but provided material that encouraged additional research and connected ongoing work to the images provided by Gemini astronauts.[20]

A conference at the end of Project Gemini featured the research of scientists and Department of Defense staff interested in photographic results, though it also included contributions and conclusions from those who worked on spacecraft engineering and astronaut experiences. Photography dominated the experiments section, with an illustrated portion written by Richard Underwood that proclaimed its benefits. He demonstrated the utility of the terrain and weather experiments with numerous photographic examples and also discussed the near-object experiments having spacecraft and astronaut subjects.[21] Descriptive narratives on the scientific results were given by Jocelyn Gill from the Office of Space Science and Applications, and Willis Foster, the director of Manned Flight Experiments. They explain how specific Gemini images showed geological features ripe for future terrestrial and orbital research along the Baja California Peninsula and Red Sea.[22] The successful performance of such experiments, they concluded, was possible because of rigorous training and procedure development akin to that used for spacecraft operations. While training for photographic work was unlikely to rival such operations, the community of scientists and engineers responsible for formulating experiments, preparing equipment, and developing procedures and training for astronauts made their satisfaction with astronaut work known through conference presentations and publications in journals and NASA books.

In addition to discussing the images at professional scientific and engineering conferences and in publications, NASA also assembled large volumes of selected images from Gemini missions issued by the Government Printing Office. Compiled by the Scientific and Technical Information Office at NASA Headquarters, the volume for Gemini missions III to V served as a conduit between NASA and the scientific community. The introduction opens with the statement that "The purpose of this volume is to provide examples of the photographs obtained from the first three manned Gemini flights, and to make them available to scientific users in various disciplines."[23] For seven dollars, anyone could purchase the book to view 242 printed color photographs (three from Gemini 3, 95 from Gemini IV, and 144 from Gemini V) selected by a panel of photographic experts mostly inside NASA. Editorial responsibility for the volume fell to Jocelyn Gill. Most captions are straightforward geographical information, indicating the utility of Richard Underwood's identification work, with some minor meteorological and geological explanations. That publication style left images wide open to scientific interpretation, and only guided casual viewers by identifying noteworthy landmarks. As scientists selected the images from mission indices, printed in lists at the end of the volume, the book had a scientific intent but left open the possibility that nonprofessionals could find value in owning the book. With this accessible format, NASA paved the way for commercially viable editions published for later missions.

Scientists interested in Gemini photography for astronomical, geographical, geological, and other scientific analysis could access the images by becoming involved in experiments, making direct requests to NASA, or reviewing them in publications. The utility of the images in their work and later publication in academic and professional publications illustrates the contemporary importance of human observations of the Earth, the Moon, and other celestial phenomena. Notable in such articles is the inclusion of photographic reproductions. As tools of science, showing the evidence that informed analysis was crucial in attempting to convince readers of the veracity of the argument. By including the photographs, crediting the photographers (either NASA or the astronaut by name), and noting processing of the frames for specific areas of interest, researchers acknowledged the necessity of visual evidence in proving their hypothesis. Based on the expectation that scientists would use such images, government scientists prepared some preliminary analysis of their own regarding the possible uses of Gemini photographs. The U.S. Geological Survey issued an early summary of images in a March 1967 report done at NASA's request, assessing the cartographic potential of orbital photography using just four images and a map based on the fourth to demonstrate their point.[24] This technical report predicted the addition of the reseau plate to Apollo Hasselblad cameras, noting that photogrammetric data from fiduciary marks to enable measurements for accurate maps.

Just one month after that report, Paul Lowman, a scientist at the Goddard Space Flight Center in Maryland, followed that initial report with his own and included seventeen photographic plates as examples.[25] This more thorough examination of Gemini photographs as material for geological research presented a preliminary overview of the value of images in research, but also some of the NASA technical information on cameras used, plus cross-references to existing works. While including NASA scientists' research, these government publications became a platform for image analysis by other scientists, laying out all of the technical information otherwise required in scientific work for creating baseline metadata. When not directly involved in shaping photographic results, researchers needed NASA to provide data after the fact to make use of astronaut images.

Later in 1967, academic journals began featuring articles that included image analysis. In the *Astrophysical Journal*, scientists from the Yerkes Observatory and Northwestern University published an article on a nebula as seen in a single Gemini XI photograph reproduced in the article.[26] They included detailed descriptions of the camera system used (the Maurer 70mm camera), wavelengths of light analyzed, and film processing for their research. While they made frequent references to the astronaut photographers, it is clear that the authors understood the benefits and drawbacks of having human operators at the controls. There was a presumption and expectation in these reports of acquiring better photographic data in the future through remotely operated telescopes and satellites. Proportionally, however, terrestrial and lunar surface image experiments received more attention than astronomical phenomena, and therefore saw wider distribution and publication in scholarly work.

Following the end of the Gemini program, even after the orbital and lunar missions of Apollo, geologists used Gemini photographs for their research into terrestrial structures over wide areas. As Lowman explained in a June 1969 article, one of the key benefits of Gemini orbital photography was the expanse of land seen in each frame, the availability of color and multispectral coverage, and the unlimited ability to disseminate the photographs.[27] Like his earlier summary report on Gemini photography, Lowman noted the many advantages and disadvantages of orbital photography for scientific work, many of which existed regardless of human or robotic involvement in image capture. While technology and uncontrollable environmental conditions limited the quality of returned images, Lowman encouraged further use of astronaut photography for mapping, tectonic and sedimentation studies, and the planning of fieldwork and regional surveys. Lowman even dedicated a short section to the "unlimited dissemination" possible because of the Space Act requirements, making the work a real boon to scientists around the world.[28] In the same publication of *Photogrammetria*, H. E. C. van der Meer Mohr similarly assessed the value of Gemini photography, although his positive commentary looks at the potential of imaging for geological mapping.[29] NASA scientists and principal investigators promoted these resources in professional journal publications, encouraging additional and nuanced research plans for later missions.

Photographic targets for specific geological areas, discussed in other articles, stemmed seemingly from Lowman and the NASA team's knowledge of current interests while planning astronaut schedules. Candidate locations required prior identification to simplify mission plans, essentially conforming photography into scripted mission documents. A brief survey of articles resulting from Gemini photographic experiments shows a confluence of research for scientific knowledge and commercial interests.[30] Geologist Frank Wobber, working for IBM and later EarthSat, published articles following both Gemini and Apollo missions that, at their core, highlighted the Earth resources knowledge gained from astronaut and other orbital photography. His heavily illustrated work frequently included notations inscribed on the images to illustrate features of interest for locating natural resources. For example, structures in and around the Red Sea were studied because of visible geological patterns that could indicate possible oil and other natural resource deposits.[31] The long-term study of this region relied heavily on astronaut photography for visual investigations until Earth resources satellites provided automated imagery collection in the early 1970s.

More important than the articles themselves to understanding the impact of astronaut-captured photographs was their use within publications. In most cases, the articles included at least a single visual reference, and often used multiple photographs, some heavily modified with cropping and pre-Photoshop hand editing (Figure 4.4). Embedded within often-lengthy textual explanations of their findings, the photographs, in their original, edited, and retouched formats, were used to support interpretations of geological, meteorological, and astronomical features and phenomena. When used as evidence, astronaut photographs not only fulfilled

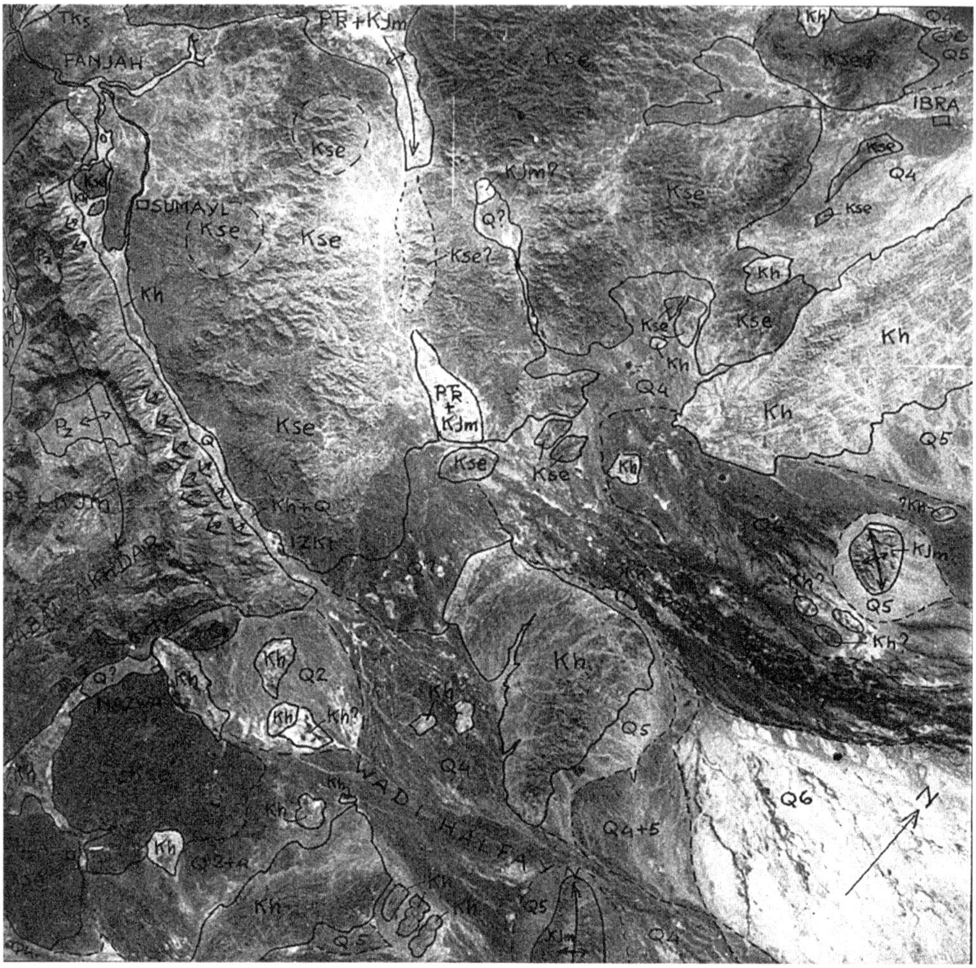

Figure 4.4. Gemini IV photograph of part of the Sultanate of Muscat and Oman, with geological interpretation, from S65-34661 (NASA).

the original goals of scientists whose photographic experiments became part of human missions, but also the needs of scientists globally, whose research interests dovetailed with the results. The photographs, as indicated by Wobber's article on planning for future satellite missions, also indicated what more frequent robotic imaging could offer to researchers. These photographs did far more than expected as a means of informing future scientific work, and engineers and scientists adjusted mission plans to support more imaging during Gemini and Apollo.

Editing photographs—tweaking them with overlaid interpretive elements or cropping them to focus on distinct features—requires examination as part of the process of dissemination because it shaped and altered the experience of the

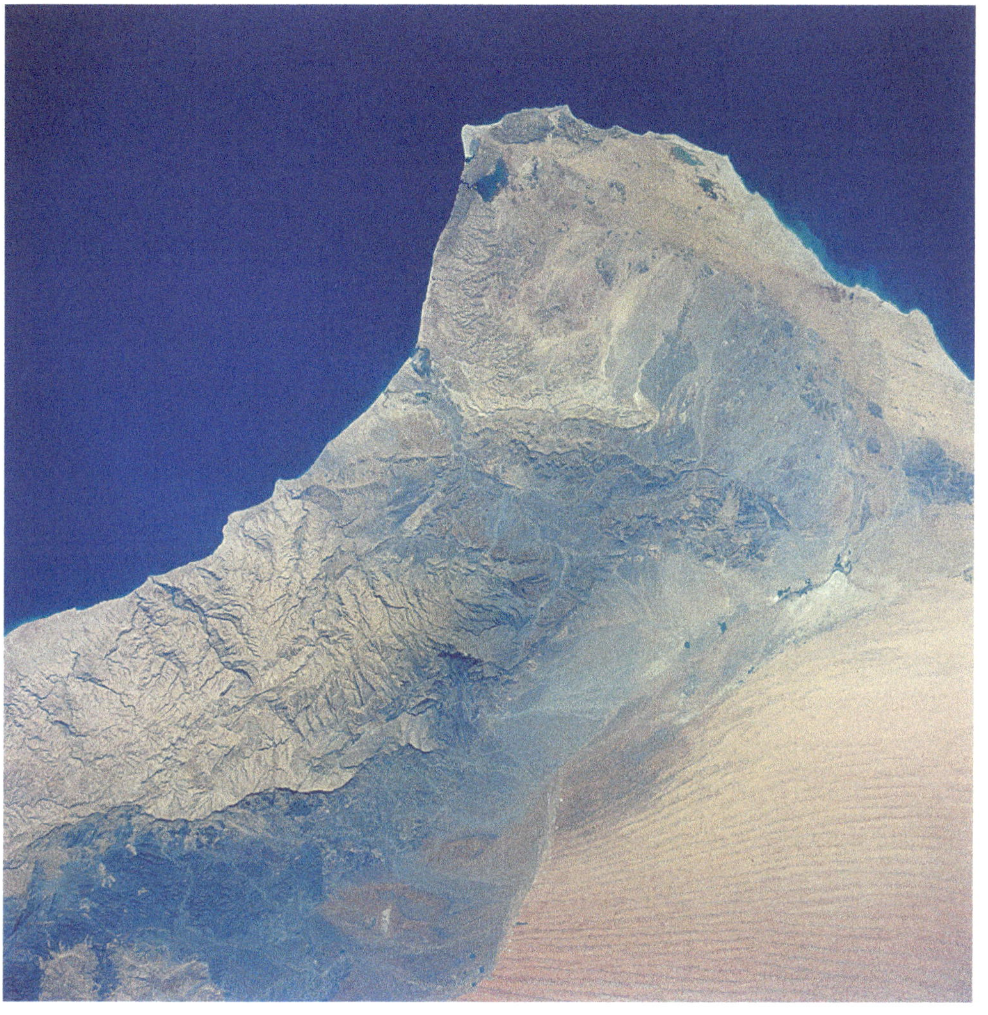

Figure 4.5. Sultanate of Muscat and Oman, Ras Al Hadd, S65-34661 (NASA).

viewers. For the audience of these articles—other scientists—only the segment of the photograph relevant to the research mattered as long at the edited material did not negate the use of the photograph as evidence. In the example from the van der Meer Mohr article, the cropped image he used as his first example is a nearly indistinguishable part of the actual photograph taken by astronauts White and McDivitt (Figure 4.5). By cropping out the vast majority of the image, the author eliminated nearly all context unless the viewer was intimately aware of the geography featured, any aesthetic value wiped out in favor of establishing the image as one of science. What this demonstrated and did repeatedly throughout the space program was that in using publically available images generated by a government source, audiences had the ability to imagine and reimagine astronaut photographs in thousands of ways. NASA managers pinpointed only a handful of those images as useful from their first release.

The exponentially higher number of images from Apollo meant that more scientists wrote even more articles on features seen in photographs. In the case of lunar missions, geological and geographical investigations surpassed all others types of research. In these specialties, physical evidence in the form of rocks and soil returned by astronauts added new dimensions, allowing scientists to enhance findings gained from photographic and physical evidence.[32] Principal investigators for imaging experiments published their findings, typically through reports prepared for or through NASA. The most prolific source of articles in the immediate aftermath of the missions came from the Apollo Lunar Geological Team. This research component of NASA was home to geologists and other scientists intimately familiar with the expectations for image and sample returns from the lunar landings—they directly shaped results by having representatives at mission planning meetings. Their research interests shaped requests for specific images and samples. As the primary audience for images, scientists were anything but disappointed by the number of film frames featuring the lunar surface and hundreds of pounds of samples returned.

Special reports by NASA scientists often served as the basis for additional work by those in the next tier of researchers, cited repeatedly in later research. With a plethora of evidence at hand in the form of images and samples, thousands of articles, conference proceedings, special journal editions, and edited volumes appeared in the decades since those missions. Even today, scientists visit Regional Planetary Image Facilities to work with first-generation masters made from the original flight films created from Apollo photographs. Samples are stored at the Lunar Receiving Lab (LRL) in Houston and are still accessible to scientists.[33] Photographs and evidence collected during Mercury, Gemini, and Apollo fulfilled the needs of those who established requirements from the start.

Outside of the scientific and technical communities, astronaut images appealed to managers and political entities responsible for making human spaceflight possible. These images took center stage next to witness testimony at the regular and special hearings in front of Congress. As with any government agency, upper-level managers worked through PAO offices to carry their message to supporters. They were also required to report regularly to Congress on their progress and any problems encountered. The most notable and publicized hearings were in the instances of the two major disasters NASA experienced during this period, the Apollo 1 fire and the Apollo 13 mission failure. NASA's use of images and their appearance in printed hearing records varied by the setting, in this case the legislative house in which the hearing took place, and the nature of record keeping for the hearings. The hearings following Apollo 13 make for an interesting case study since the Senate record includes a handful of drawings, but no photographs of parts or other engineering information specific to the accident.[34] On the other hand, the House hearings were extensive and included a copy of the accident review board's report. Within the live House hearings, NASA representatives showed slides of the equipment in question as photographed prior to the flight. Astronaut photographs only served as supporting evidence within the context of the review board's supplemental material, showing the service module damage caused by the oxygen tank explosion, printed as an appendix in the final record of the hearings.[35] Perhaps ironically, photography of the actual failed technologies played almost no role in explaining the problem to nonspecialists for this "successful failure," though the event itself is perhaps one of the program's most well-known now thanks to Hollywood's depiction of the mission.

Throughout its history, NASA submitted biannual reports to Congress to explain their activities over the previous half year. While not a comprehensive review of the work done by astronauts, the report encompassed the first two crewed Apollo flights, included a number of photographs and other illustrations, and material on the robotic spaceflight activities of NASA work.[36] Annual authorization hearings to renew NASA's mission as an agency and establish funding needs became another venue to present visual evidence from astronaut experiences, typically at the beginning of the slides to grab attention. For example, during the fiscal year (FY) 1971 hearing, NASA managers presented a few astronaut photographs, but stepped up the presentation in their FY1972 hearing in early 1971 with a video of Apollo 14 mission highlights narrated in person by the crew. They continued in the hearing by speaking about their experiences while showing slides of the photographs taken on the Moon.[37] Both Apollo 15 and 16 crews presented testimony and photographs during informational hearings on their missions, while the Apollo 17 narrative became part of the FY1974 authorization hearing that featured only geologist-astronaut Harrison Schmitt.[38] What I offer here is only a summary of the late Apollo program Congressional hearings and the use of photographs, but a comprehensive examination of NASA preparations for such hearings may tell a more complex story. Additional research may also illuminate what responses NASA expected to the inclusion of images or how persuasive visual evidence may have ultimately been in accomplishing the funding goals of these hearings.

For the professional science and political communities, astronaut photography more than fulfilled expectations. Scientists and legislative affairs staff at NASA had thousands of images from which to work when looking to explain lunar features at hearings on Capitol Hill. The scientific community's objective standards for evidence made the use of astronaut photographs far more persuasive and significant than how NASA used images in the highly subjective setting of Congressional hearings, which depended on a far more diverse and intricate set of evidence than photographs or research alone.

Non-Professionals

Within the public context of publication, photographs taken by astronauts found an undoubtedly murkier and complex response. People outside the primary audiences for astronaut photographs had fewer technical but more thoughtful and emotional reactions to the images. It is interesting to compare spaceflight visuals released following the Apollo missions and those from other news stories competing for the same space in the media. One of the most visually disturbing events of the period was the Vietnam War. Nightly news coverage regularly brought a national audience compelling, troubling images of the successes and failures of the American military forces in Southeast Asia. As Rob Kroes suggested regarding iconic images of the Vietnam War by Eddie Adams and Nick Ut, it is possible for still photographs to possess a concentrated power thanks to their circulation via mass media, which I would argue is true about astronaut images as well.[39] Also helpful to understanding the connection between memory and images of the Vietnam War is the first chapter of Marita Sturken's *Tangled Memories.*[40] She explains that one agreed upon point about the war was that it was divisive, but that nearly all photographs in the public sphere supported the understanding of it as a brutal attack on the innocent people of Vietnam. These confusing and conflicting visuals battled for space in print and on television, but scholarly analysis of images broadly as mnemonic devices aides my argument on the meaning of astronaut photography in American culture.

On a less frequent basis, usually about three times each year, newspapers and television programs provided viewers with images of active and spectacular space missions that sent humans around and to the surface of the Moon.[41] It came down to producers of printed and television news outlets to decide if and how they would allocate space and time to these events. The public's interaction with and memories of these historic events was primarily a mediated one, designed by a combination of those taking the images, those releasing the images, and those making the choices on what made it in front of the public. The politics of the printed news publication process had tremendous impact on the reception of astronaut photography by the reading public.

Before examining the ways in which the public encountered astronaut photographs in traditional print media, another form of direct public engagement by NASA requires consideration. NASA managers were well aware of their federal mandate to explain their actions to the public.[42] Visual information required interpretation, so for previously published volumes for Gemini, simple identification of geological features by Richard Underwood were sufficient.[43] The success of those books among scientists and the public prompted internal NASA discussion of similar Apollo books after the successful return of Apollo 8 from the Moon.

In March 1969, memos circulated between NASA managers about compiling similar books to highlight the photographic achievements of astronauts during Apollo. Unlike the simple layout and scientific focus of Gemini photographic volumes, Apollo publications took a different form and sought a different audience. In planning for Apollo photographic books, NASA expected far more photography to sift through, which necessitated consideration for how to best use the resource. Sales expectations were also an issue, and managers set sights very high. Writing to MSC Director Robert Gilruth, LRC (Langley Research Center) Director Edgar Cortwright encouraged the compilation of a book similar to those published during Gemini, in part to appeal as a Christmas gift if the book was finished in time for the 1969 season. According to his memo, sales of the Gemini book associated with photographs from the first three missions sold nearly 30,000 copies after its 1966 release.[44] Additional encouragement came in this period with news that *Exploring Space With a Camera*, the 1968 publication edited by Cortwright, which included only a small portion on astronaut photography, sold out from the Government Printing Office at 54,000 copies.[45] The positive attention yielded by such volumes quickly became an increasing part of NASA discussions.

In a quick succession of memoranda in March and April 1969, MSC and headquarters staff agreed as to the desirability of a publication for Apollo images. Laid out as a simple index, the workload fell to the PTL staff to assemble images and develop text with support from a variety of departments and the top management at MSC.[46] Teams developed these special publications to reflect the number of images collected, but the quantity and quality of the products were inconsistent generally when examined across the program. Surveying the list of official reports shows only Apollos 8, 10, and 12 had mission-specific photographic reports in this series. Following the Apollo 13 mission, photographic reports were integrated into larger mission reports on either science or the mission itself, ending the index format for printing results. Even so, astronaut photography easily served a reference function in reports, giving NASA many opportunities to feature this unique resource.

Newspapers and magazines were certainly the most accessible public means for viewing astronaut-captured images. Simple captions accompanied the stories, and only occasionally did publications carry extensive descriptions to explain the content and context of the photograph. Because these pictures appeared set near or within the text of other stories selected by editors, it is important to analyze the space program as part of the larger cultural landscape of the late 1960s in which NASA's human spaceflight program occurred. Because of the frequency of human missions in this period, case studies from the start and end of Apollo help us understand how at both the height of popularity and end of the era, people encountered astronaut photographs in their daily lives.

Exploring the politics behind news publishing both inside and between outlets, as well as the relationship between news publishers and NASA staff, offers a means to understand the importance of astronaut images to print media. These relationships were refined regularly and influenced how and where photographs appeared, particularly when placement on the front page was possible. The 1960s was a period of massive expansion of the representation of science and technology reporters in newsrooms, which scholars suggest was a visible reaction to Sputnik and the race to the Moon.[47] How readers encountered astronaut photography became a seamless part of their consumption of the most salient current events of the day. This negotiated situation, invisible to readers, involved writing and organizing news in ways editors and publishers felt would sell more issues, garner editorial respect within the community of news outlets and critics, and uphold principles held by the publishers and editors of the papers.[48] While the statistical popularity of the space program at this time was less than legend might indicate now, the most widely published newspapers and magazines of the 1960s offered brief and memorable glimpses of life in space for readers. That exposure to astronaut photography carried NASA's narrative more broadly and for longer than anyone expected.

The media played an important role in shaping public understanding of NASA's human spaceflight program from even before the 1959 announcement of the first group of seven astronauts. Enthusiasm for space topics from films to television to artwork and other visual forms permeated postwar American culture.[49] These imaginative creations, sometimes informed by rocket engineers, laid the groundwork for expectations of actual spaceflight. The news media became the conduit by which readers came to understand how space appeared and felt, with reporters and publishers translating the words and images of astronauts into digestible, understandable, and relatable stories. This put print media outlets in an interesting position over a decade known for a dramatic rise in the ownership and popularity of televisions, intense televised events such as the assassination of President Kennedy, and the war in Vietnam. Former *Houston Chronicle* Science and Space Editor Warren Burkett wrote in his instructional book for up-and-coming science reporters that the primary criteria for reaching audiences through print are "timeliness, timing, impact, significance, uniqueness, and human interest."[50] In an industry dependent on placing appealing stories on the front page to increase readers, newspaper and magazine reporters needed to understand their audience in order to satisfy publishers whose profits came from advertising and increasing readership.[51] The unique appeal of human spaceflight encouraged media outlets to keep it and the experiences of astronauts on the front page since it was their best option for grabbing attention about the events television networks carried live. While timelines may have been on the side of the television stations, the lasting impressions made by printed images meant print media outlets had a more serious and permanent role in forming the visual memories for the public.[52]

Print media coverage of Apollo 8 provides ample material for analyzing the broad coverage of a single event, and how that reflected the newsroom and public debates about the importance of spaceflight. Enthusiasm for the unique and exciting topic of the first human lunar flight made excellent fodder for the editors of papers and magazines, most interested in selling as many copies as possible while conveying compelling stories to encourage repeat readers. Knowing their audience meant knowing that not every copy would be snatched up based on putting astronauts on the front page, but the tremendous reach of print media meant impressions using astronaut photography had wide viewing among the U.S. population.[53] Like any other topic, NASA missions competed for position in publications, but as a subject of the science and technology reporting, there was additional pressure on reporters to energize stories and bring their best to the newsroom in order for their material to make the front page.[54] If we agree with the assertion that the most important dynamic in creating news is momentum, the space program built that in spades over the first decade of NASA's human spaceflight program.

The impact of publishing *Earthrise* within the complicated period of late 1968 and early 1969 publications shows the incredible popularity of the space program as reflected in major newspapers across the U.S. despite national and international problems of the moment. For example, the *New York Times* published a horizontally oriented version of the color *Earthrise* photograph from Apollo 8 on the front page of their December 30, 1968, edition with an article by writer Homer Bigart.[55] Nearby "above the fold" stories include reports of New York state budget needs and the concerns of the Johnson administration and United Nations about Israel's raid on Beirut the previous day (Figure 4.6). On the same day, the *Los Angeles Times* used a similarly oriented version of Anders' color photograph on their front page, though the headline of the day regarding the attack on Beirut dominates, confusing the notion of which story the paper considered to be the most newsworthy (Figure 4.7). Important to note here is that newspapers of the time only printed special sections, usually the comics, in color, so despite having a color image on hand it was inevitably read as black and white. These two instances not only provide a sense of the ways in which one of the most historic moments appeared in printed media, but also how that moment competed for coverage with other news events.

Other publications of *Earthrise* show the varying context of its initial printing in different newspapers. The *Washington Post* used a heavily cropped version of *Earthrise* on the last day of 1968, though it played second fiddle on the front page to a solitary Earth image from the translunar portion of the journey. Without an accompanying story, the textual surroundings for the images are articles about Lebanon's response to Israeli attacks, District of Columbia city council members seeking to overturn mayoral decisions on police use of guns, and Senator Edward Kennedy's challenge for the majority whip job. While the Apollo 8 images dominated front pages of some major publications, the surrounding text focused on issues pertaining to local politics (Figure 4.8). While sometimes portrayed today as an

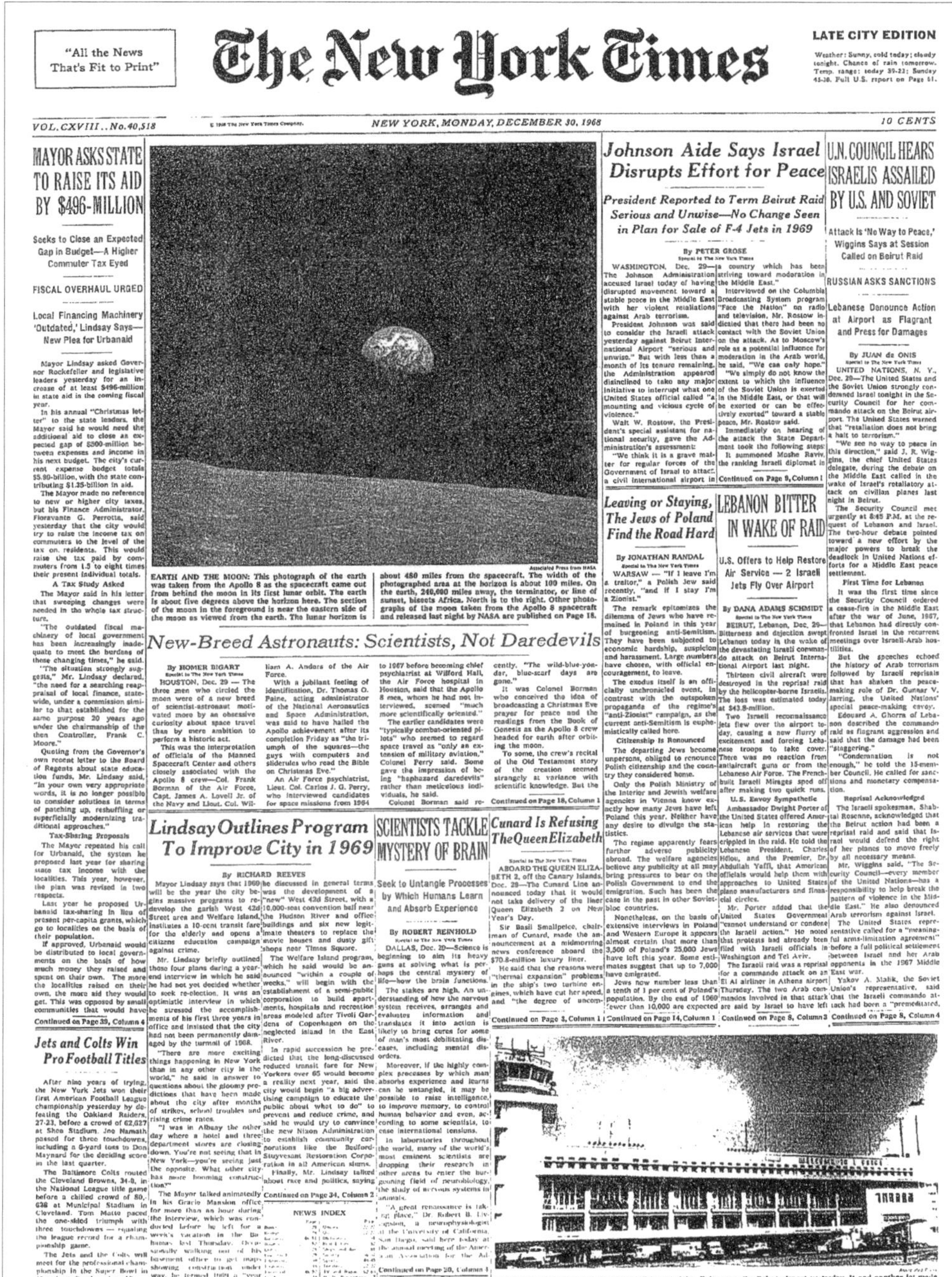

"All the News That's Fit to Print"

The New York Times

LATE CITY EDITION

VOL. CXVIII..No. 40,518 — NEW YORK, MONDAY, DECEMBER 30, 1968 — 10 CENTS

MAYOR ASKS STATE TO RAISE ITS AID BY $496-MILLION

Seeks to Close an Expected Gap in Budget—A Higher Commuter Tax Eyed

FISCAL OVERHAUL URGED

Local Financing Machinery 'Outdated,' Lindsay Says—New Plea for Urbanaid

EARTH AND THE MOON: This photograph of the earth was taken from the Apollo 8 as the spacecraft came out from behind the moon in its first lunar orbit. The earth is about five degrees above the horizon here. The section of the moon in the foreground is near the eastern side of the moon as viewed from the earth. The lunar horizon is about 480 miles from the spacecraft. The width of the photographed area at the horizon is about 100 miles. On the earth, 240,000 miles away, the terminator, or line of sunset, bisects Africa. North is to the right. Other photographs of the moon taken from the Apollo 8 spacecraft and released last night by NASA are published on Page 18.

Johnson Aide Says Israel Disrupts Effort for Peace

President Reported to Term Beirut Raid Serious and Unwise—No Change Seen in Plan for Sale of F-4 Jets in 1969

U.N. COUNCIL HEARS ISRAELIS ASSAILED BY U.S. AND SOVIET

Attack Is 'No Way to Peace,' Wiggins Says at Session Called on Beirut Raid

RUSSIAN ASKS SANCTIONS

Lebanese Denounce Action at Airport as Flagrant and Press for Damages

New-Breed Astronauts: Scientists, Not Daredevils

Leaving or Staying, The Jews of Poland Find the Road Hard

LEBANON BITTER IN WAKE OF RAID

U.S. Offers to Help Restore Air Service — 2 Israeli Jets Fly Over Airport

Lindsay Outlines Program To Improve City in 1969

SCIENTISTS TACKLE MYSTERY OF BRAIN

Seek to Untangle Processes by Which Humans Learn and Absorb Experience

Cunard Is Refusing The Queen Elizabeth

Jets and Colts Win Pro Football Titles

ISRAELI RECONNAISSANCE: An Israeli jet, at right, flying over the Beirut airport yesterday. It and another jet made two passes over the scene of Saturday's helicopter-borne commando raid and left without firing or being fired upon.

Figure 4.6. *New York Times* front page, December 30, 1968 (From the *New York Times*. Associated Press).

Astronauts View Moon Photos

MONDAY

COMPLETE RACING

Los Angeles Times

FINAL

LARGEST CIRCULATION IN THE WEST, 958,124 DAILY, 1,269,931 SUNDAY.

VOL. LXXXVIII FIVE PARTS—PART ONE MONDAY MORNING, DECEMBER 30, 1968 90 PAGES Copyright © 1968 Los Angeles Times DAILY 10c

U.S. Urges Israel:

'REGRET' AIR RAID

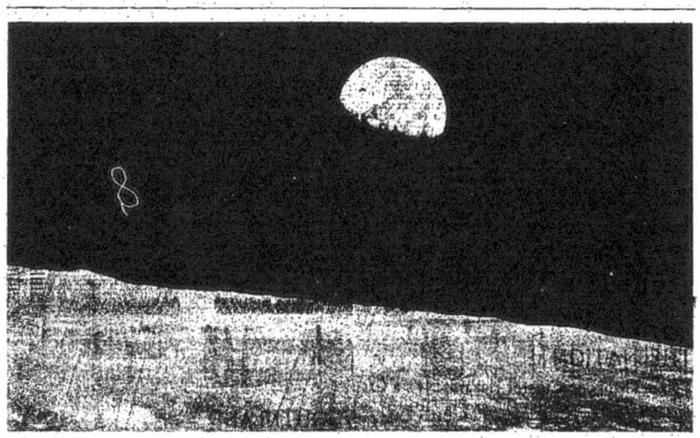

AS APOLLO 8 SAW THE GOOD EARTH—Picture brought back from space shows the earth rising as astronauts come from behind the moon. Line between darkness and light on earth bisects Africa. The area in foreground is near eastern side of moon as seen from earth. NASA photo via AP Wirephoto

Lebanon Asks U.N. Sanctions, Full Compensation for Damage

BY EARL W. FOELL
Times Staff Writer

UNITED NATIONS—The United States urged Israel Sunday night to make "a candid expression of regret" for its commando raid Saturday on Beirut International Airport. American Ambassador J. R. Wiggins issued this plea for "an honest acknowledgement of misgivings" during the opening stages of a U.N. Security Council debate.

He said that "surely Israel must be having sober second thoughts" in light of international reactions to the raid.

Lebanon led off the late night session by demanding that the Security Council vote sanctions against Israel and arrange "full compensation" for an asserted $50 million or more of damage to airliners at Beirut airport.

Israel countered by charging the Lebanese government with failure to "disband terror organizations" operating in Lebanon. The Popular Front for the Liberation of Palestine issued a statement in Beirut saying its members carried out Thursday's two-man attack on an El Al airliner at Athens airport.

Asks For Council Action

Israeli delegate Shabtai Rosenne asked the council to show it would "no longer tolerate" terrorist warfare from Arab soil.

Delegates of major powers on the Security Council said privately that they doubted Lebanon could win either sanctions or compensation. They also doubted that Israel could avoid U.N. condemnation for its raid on an international civilian airport.

Wiggins indicated that such condemnation was in store by stating flatly that Washington would support "prompt action to condemn" this latest Israeli retaliation raid.

He did so despite earlier expressing American sympathy with the outrage that Israelis had felt after the two-man terrorist attack at Athens airport.

"In magnitude," Wiggins concluded, the Israeli retaliation at Beirut was "entirely disproportionate." He warned, speaking for Washington, that "retaliation does not bring an end to terrorism" but instead expands it.

American diplomats here have recently expressed private concern that the pattern of Israeli retaliations tends to enlarge Israel's circle of active opponents. They cite the recent raid against an Iraqi unit in Jordan as well as the helicopter-borne commando attack in Beirut.

One American delegate said Sunday night that he could see Israel's psychology in trying "to make the cost five times the entry fee for Arab bystanders who try to prove their devotion by loudly championing terrorist groups." But, he added, "I think Israel is making real military opponents out of what were only barroom talkers before the retaliatory raids."

Soviet Ambassador Jacob A. Malik condemned Israel and blamed the United States for encouraging Israel. Malik said Wiggins had taken a new turn in U.S. policy by condemning Israel. Malik said the Security Council should require Israel to punish those who took part in the Beirut attack.

Please Turn to Page 7, Col. 1

ISRAELI JETS BUZZ BEIRUT'S AIRPORT, STIR FRESH PANIC

BEIRUT (UPI)—Widely condemned for the commando attack that destroyed the bulk of Lebanon's civil airlines fleet, Israel sent two jets on reconnaissance flights Sunday over the still smoldering wreckage at Beirut's International Airport.

Low passes by the scouting planes stirred panic at the airport for the second time in 12 hours. Lebanese troops dived for cover. There was no reaction from antiaircraft guns or from Lebanese fighters, based 40 miles away at Rawak. The Israeli jets sped off after making two quick runs.

A British Overseas Airways plane had taken off and passengers were checking in for a Pan American World Airways and a Dutch KLM flight when the Israeli pilots showed up to assess the damage wrought in the lightning raid Saturday night.

Thirteen planes lay demolished.

These were three Comets, two Caravelles, a Boeing 707, a VC-10

Please Turn to Page 6, Col. 5

New Storm May Bring More Rain on New Year's Day

The edge of a storm front moved toward Southern California Sunday and forecasters said it may bring occasional rain to dampen New Year's celebrations Wednesday.

Clearing weather is expected later

Storm	.02
Season to date	2.33
Last year this date	11.33
Normal	4.43

New Year's Day but there is a chance that "left over" showers may mar the Tournament of Roses Parade.

However, the weather should be clear, brisk and windy by kickoff time for the Rose Bowl game in the afternoon, a Universal Weather Service forecaster said.

The threatening front is from a storm which was centered Sunday about 400 miles off the Washington-Oregon coast. The main storm itself is expected to swing eastward and pass through the Seattle area. But a weakening frontal system will push farther south, passing through the Los Angeles basin shortly after midnight Tuesday, forecasters said.

The pattern of the frontal movement could be similar to one which moved into Southern California late Saturday, leaving only .02 inch of

Please Turn to Page 27, Col. 8

THE WEATHER

Mostly clear today. Occasional rain late Tuesday night. High today and Tuesday near 64. High Sunday, 61; low, 50.

Complete weather information on Page 6, Part 3.

OCCUPATION IS KEY

Allies Adopt Hard Attitude in Pacifying S. Vietnam Areas

BY JACK FOISIE
Times Staff Writer

SAIGON—A noticeable change has been made in the pacification program in South Vietnam.

The change is in the initial step: Instead of holding out the olive branch as government troops enter a known or suspected Viet Cong village, the troops now occupy the hamlet.

And the occupiers make no bones about the fact they are there to stay. Friendship and material assistance can follow if the people mind their ways. But any inner resistance or any surreptitious dealings with the enemy in the next rice paddy, if discovered, will mean punishment.

This is not exactly how the high-level administrators of the program would say it works. But that is the way the combined forces—American and Vietnamese—are working it in northernmost Quang Tri province.

And according to Marine Maj. Gen. Raymond G. Davis Jr. and others in the province who are conducting this first phase of pacification, it seems to be the way to that long-sought objective: "Winning the hearts and minds" of the peasants and gaining their loyalty to the Saigon government.

If true, it has been a long time coming.

The American mission in Vietnam is now on its fifth or so pacification program. The previous programs were all failures to varying degrees. In 1963, under American-backed President Ngo Dinh Diem, there was the Sunrise Village Program. Next Strategic Hamlets, followed by the Oil Spot Theory, the New Life Hamlet, Revolutionary Development, and others.

All sorts of organizations were

Please Turn to Page 13, Col. 1

Index to The Times

Astronauts Spend Time With Families, View Moon Photos

HOUSTON (UPI) — America's moon-orbiting Apollo 8 astronauts received a tumultuous welcome home early Sunday, spent a few hours with their families and then took a look at the spectacular photographs they shot of the lunar surface.

Only a few hours after Frank Borman, James A. Lovell Jr. and William A. Anders glimpsed their own handiwork, the space agency released the photographs to the public.

In color prints, the moon's surface appeared to be yellowish-gray, just like "wet beach sand," the astronauts described during their epic adventure to the moon and back. The moon's craters were sharply defined in the photographs.

Gemini 7 Photo Reversed

One shot showed the earth far away in the background and the moon in the foreground.

"Remember the shot from Gemini 7 (which Borman flew in 1965) with the earth in the foreground and the moon in the background," said Paul Haney, chief spokesman for the manned spacecraft center. "Well, here's the reverse."

The earth hung like a blue ball in the background of this picture, about the size of a quarter, its brownish continents almost obscured by fluffy white clouds.

A color shot showed the crater Longrenus, 80 miles in diameter, and located just east of the Sea of Fertility on the right hand edge of the moon's face. This was one of the

Please Turn to Page 20, Col. 1

NLF 'Liberation Committees' Held Part of Complex Strategy

BY ARTHUR J. DOMMEN
Times Staff Writer

PARIS—The National Liberation Front apparently is aiming at a series of objectives in both Paris and South Vietnam with the "liberation committees" and "liberation councils" it asserts it has established through elections in much of Vietnam, observers here believe.

The NLF purports to be the authentic representative of the South Vietnamese people, and observers see the reports of widespread grass-roots organizations as designed partly to bolster the NLF's standing at the Paris talks.

By setting up the committees in wartime, the NLF can hope to reap the political benefits of any settlement. When the war ends, the elected officials will be able to claim a share of the credit for peace and the withdrawal of U.S. troops in the same way the Viet Minh took the political credit for forcing French withdrawal.

Rather than challenging the Saigon government in the immediate future, the "liberation committees" probably are intended for a second-phase mission—preparing for and exploiting general elections at some future time, observers believe. The committee, probably with a backbone of NLF cadres, will organize voting in favor of the NLF.

There is risk of setting up the committees before a negotiated set-

Please Turn to Page 11, Col. 1

Figure 4.7. *Los Angeles Times* front page, December 30, 1968 (Copyright © 1968, *Los Angeles Times*. Used with Permission).

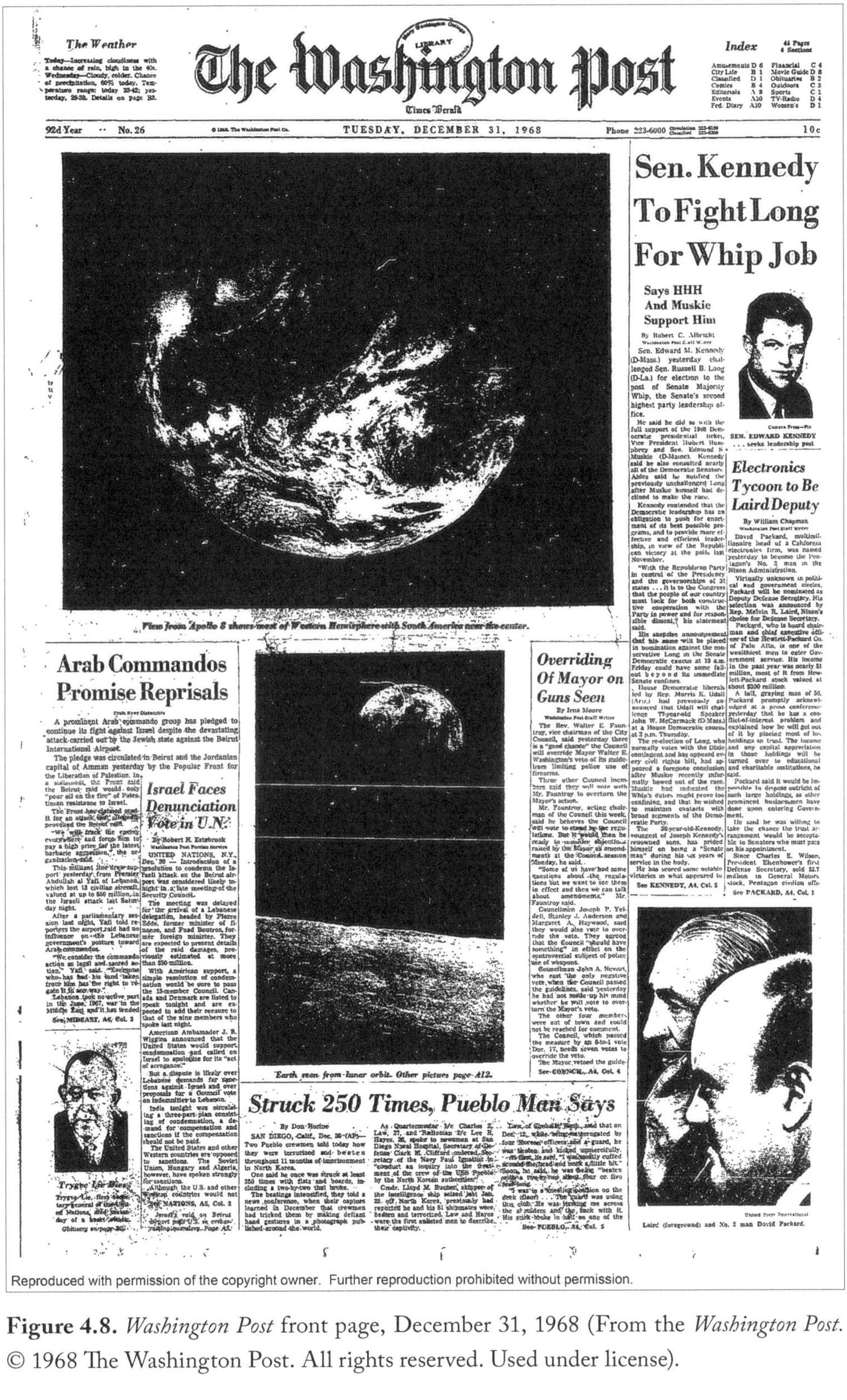

The Weather

Today—Increasing cloudiness with a chance of rain, high in the 40s. Wednesday—Cloudy, colder. Chance of precipitation, 60% today. Temperature range: today 33-42; yesterday, 29-32. Details on page B3.

The Washington Post

Times Herald

Index — 44 Pages, 4 Sections

Amusements	D 6	Financial	C 4
City Life	B 1	Movie Guide	D 8
Classified	D 1	Obituaries	B 2
Comics	B 4	Outdoors	C 3
Editorials	A 8	Sports	C 1
Events	A10	TV-Radio	D 4
Fed. Diary	A10	Women's	D 1

92d Year · No. 26 · © 1968, The Washington Post Co. · TUESDAY, DECEMBER 31, 1968 · Phone 223-6000 · 10c

View from Apollo 8 shows most of Western Hemisphere with South America near the center.

Sen. Kennedy To Fight Long For Whip Job

Says HHH And Muskie Support Him

By Robert C. Albright
Washington Post Staff Writer

Sen. Edward M. Kennedy (D-Mass.) yesterday challenged Sen. Russell B. Long (D-La.) for election to the post of Senate Majority Whip, the Senate's second highest party leadership office.

He said he did so with the full support of the 1968 Democratic presidential ticket, Vice President Hubert Humphrey and Sen. Edmund S. Muskie (D-Maine). Kennedy said he also consulted nearly all of the Democratic Senators. Aides said he notified the previously unchallenged Long after Muskie himself had declined to make the race.

Kennedy contended that the Democratic leadership has an obligation to push for enactment of its best possible programs, and to provide more effective and efficient leadership, in view of the Republican victory at the polls last November.

"With the Republican Party in control of the Presidency and the governorships of 31 states . . . it is to the Congress that the people of our country must look for both constructive cooperation with the Party in power and for responsible dissent," his statement said.

His surprise announcement that his name will be placed in nomination against the conservative Long in the Senate Democratic caucus at 10 a.m. Friday could have some fallout beyond its immediate Senate confines.

House Democratic liberals led by Rep. Morris K. Udall (Ariz.) had previously announced that Udall will challenge 77-year-old Speaker John W. McCormack (D-Mass.) at a House Democratic caucus at 2 p.m. Thursday.

The re-election of Long, who normally votes with the Dixie contingent and has opposed every civil rights bill, had appeared a foregone conclusion after Muskie recently informally bowed out of the race. Muskie had indicated the Whip's duties might prove too confining, and that he wished to maintain contacts with broad segments of the Democratic Party.

The 36-year-old Kennedy, youngest of Joseph Kennedy's renowned sons, has prided himself on being a "Senate man" during his six years of service in the body.

He has scored some notable victories in what appeared to

See KENNEDY, A4, Col. 5

Camera Press—Pix
SEN. EDWARD KENNEDY . . . seeks leadership post

Electronics Tycoon to Be Laird Deputy

By William Chapman
Washington Post Staff Writer

David Packard, multimillionaire head of a California electronics firm, was named yesterday to become the Pentagon's No. 2 man in the Nixon Administration.

Virtually unknown in political and government circles, Packard will be nominated as Deputy Defense Secretary. His selection was announced by Rep. Melvin R. Laird, Nixon's choice for Defense Secretary.

Packard, who is board chairman and chief executive officer of the Hewlett-Packard Co. of Palo Alto, is one of the wealthiest men to enter Government service. His income in the past year was nearly $1 million, most of it from Hewlett-Packard stock valued at about $300 million.

A tall, graying man of 56, Packard promptly acknowledged at a press conference yesterday that he has a conflict-of-interest problem and explained how he will get out of it by placing most of his holdings in trust. The income and any capital appreciation in those holdings will be turned over to educational and charitable institutions, he said.

Packard said it would be impossible to dispose outright of such large holdings, as other prominent businessmen have done upon entering Government.

He said he was willing to take the chance the trust arrangement would be acceptable to Senators who must pass on his appointment.

Since Charles E. Wilson, President Eisenhower's first Defense Secretary, sold $2.7 million in General Motors stock, Pentagon civilian offi-

See PACKARD, A4, Col. 1

Arab Commandos Promise Reprisals

From News Dispatches

A prominent Arab commando group has pledged to continue its fight against Israel despite the devastating attack carried out by the Jewish state against the Beirut International Airport.

The pledge was circulated in Beirut and the Jordanian capital of Amman yesterday by the Popular Front for the Liberation of Palestine. In a statement, the Front said the Beirut raid would only "pour oil on the fire" of Palestinian resistance to Israel.

The Front has claimed credit for an attack that allegedly provoked the Beirut raid.

"We will track the enemy everywhere and force him to pay a high price for the latest barbaric aggression," the organization said.

This militant line drew support yesterday from Premier Abdullah al Yafi of Lebanon, which lost 13 civilian aircraft, valued at up to $50 million, in the Israeli attack last Saturday night.

After a parliamentary session last night, Yafi told reporters the airport raid had no influence on the Lebanese government's posture toward Arab commandos.

"We consider the commando action as legal and sacred action," Yafi said. "Everyone who has had his land taken from him has the right to regain it in any way."

Lebanon took no active part in the June, 1967, war in the Middle East and it has tended

See MIDEAST, A6, Col. 3

Israel Faces Denunciation Vote in U.N.

By Robert H. Estabrook
Washington Post Foreign Service

UNITED NATIONS, N.Y., Dec. 30 — Introduction of a resolution to condemn the Israeli attack on the Beirut airport was considered likely tonight in a late meeting of the Security Council.

The meeting was delayed for the arrival of a Lebanese delegation, headed by Pierre Edde, former minister of finance, and Fuad Boutros, former foreign minister. They are expected to present details of the raid damages, previously estimated at more than $50 million.

With American support, a simple resolution of condemnation would be sure to pass the 15-member Council. Canada and Denmark are listed to speak tonight and are expected to add their censure to that of the nine members who spoke last night.

American Ambassador J. R. Wiggins announced that the United States would support condemnation and called on Israel to apologize for its "act of arrogance."

But a dispute is likely over Lebanese demands for sanctions against Israel and over proposals for a Council vote on indemnities to Lebanon.

India tonight was circulating a three-part plan consisting of condemnation, a demand for compensation and sanctions if the compensation should not be paid.

The United States and other Western countries are opposed to sanctions. The Soviet Union, Hungary and Algeria, however, have spoken strongly for sanctions.

Although the U.S. and other Western countries would not

See NATIONS, A5, Col. 2

Israel's raid on Beirut airport puts U.S. in embarrassing quandary. Page A5.

Trygve Lie Dies

Trygve Lie, first Secretary-general of the United Nations, died yesterday of a heart attack.

Obituary on page B6.

Earth seen from lunar orbit. Other pictures page A12.

Overriding Of Mayor on Guns Seen

By Irna Moore
Washington Post Staff Writer

The Rev. Walter E. Fauntroy, vice chairman of the City Council, said yesterday there is a "good chance" the Council will override Mayor Walter E. Washington's veto of its guidelines limiting police use of firearms.

Three other Council members said they will vote with Mr. Fauntroy to overturn the Mayor's action.

Mr. Fauntroy, acting chairman of the Council this week, said he believes the Council will vote to stand by the regulations. But it would then be ready to consider objections raised by the Mayor as amendments at the Council session Monday, he said.

"Some of us have had some questions about the regulations but we want to see them in effect and then we can talk about amendments," Mr. Fauntroy said.

Councilmen Joseph P. Yeldell, Stanley J. Anderson and Margaret A. Haywood, said they would also vote to override the veto. They agreed that the Council "should have something" in effect on the controversial subject of police use of weapons.

Councilman John A. Nevius, who cast the only negative vote when the Council passed the guidelines, said yesterday he had not made up his mind whether he will vote to overturn the Mayor's veto.

The other four members were out of town and could not be reached for comment.

The Council, which passed the measure by an 8-to-1 vote Dec. 17, needs seven votes to override the veto.

The Mayor vetoed the guide-

See COUNCIL, A4, Col. 4

Struck 250 Times, Pueblo Man Says

By Don Horine

SAN DIEGO, Calif., Dec. 30 (AP)—Two Pueblo crewmen told today how they were terrorized and beaten throughout 11 months of imprisonment in North Korea.

One said he once was struck at least 250 times with fists and boards, including a two-by-two that broke.

The beatings intensified, they told a news conference, when their captors learned in December that crewmen had tricked them by making defiant hand gestures in a photograph published around the world.

As Quartermaster 3/c Charles Z. Law, 27, and Radioman 2/c Lee R. Hayes, 26, spoke to newsmen at San Diego Naval Hospital, Secretary of Defense Clark M. Clifford ordered Secretary of the Navy Paul Ignatius to "conduct an inquiry into the treatment of the crew of the USS Pueblo by the North Korean authorities."

Cmdr. Lloyd M. Bucher, skipper of the intelligence ship seized last Jan. 23 off North Korea, previously had reported he and his 81 shipmates were beaten and terrorized. Law and Hayes were the first enlisted men to describe their captivity.

Law, of Chehalis, Wash., said that on Dec. 12, while being interrogated by four Korean officers and a guard, he was beaten and kicked unmercifully.

At first, he said, "I was mostly cuffed around the head and neck a little bit." Soon, he said, he was being "beaten with a two-by-two about four or five feet long."

"I was in a kneeling position on the deck (floor) . . . The guard was using this club. He was striking me across the shoulders and the back with it. His stick broke in half on one of the

See PUEBLO, A4, Col. 5

United Press International
Laird (foreground) and No. 2 man David Packard.

Figure 4.8. *Washington Post* front page, December 31, 1968 (From the *Washington Post*.

all-consuming public event, one characterized as having saved 1968 from the year's disastrous events, the story of the first humans flying around the Moon appeared in newspapers only minimally.

Another part of the story of disseminating images involves decision making at these publications. What appears on the front page in particular involves an entirely different set of political considerations, certainly a rich area for future research. Like the daily newspapers, news magazines such as *Time*, *Life*, and *Newsweek* spent considerable cover and interior space publishing the photographic and technological achievements of Apollo 8. Full-color spreads from issues of each magazine, *Time*'s edition of January 3, 1969, *Life*'s issue on January 11, 1969, and *Newsweek*'s printings of January 6 and July 7, 1969, all featured lengthy stories and reproductions of *Earthrise* in some form, but never as prominently as those first newspapers from just days after the mission. Considering their longer story development period, the appearance of astronaut photographs within such magazines is a testament to their lasting appeal from a narrative and commercial perspective.

Another example of a contemporary publication that discussed Apollo 8 photography, an article written anonymously for the journal *Nature*, mentions what the author thought were six of the most important images from the mission. The *Earthrise* image is not among those six.[56] This may be at least one early indication that the "special correspondent" who wrote the article, like NASA, spent little time contemplating the long-term emotional potential of an image of Earth rising from around the Moon's surface in the earliest of publications after the mission. The author focused on the scientific and engineering benefits of astronaut photography. Newspapers at least, while prominently featuring *Earthrise*, steered public interest toward the photographs, but other stories of the day provided plenty of distraction for those uninterested in the space program.

Publication of the *Whole Earth* image (also referred to as the *Blue Marble*) from Apollo 17 likewise spread a single image quickly and broadly. Papers were again constrained by the post-mission development and dissemination process at NASA, so the image only made front pages in time for Christmas Eve 1972, five days after the command module and crew landed safely in the Pacific Ocean. Contemporary articles on the topic of the image's release nearly universally proclaim its instant worldwide popularity thanks to publication on nearly every newspaper front page.[57] But while the *New York Times* featured it prominently on the front page on December 24, 1972 (Figure 4.9), and the *Boston Globe* used its entire front page for the image on Christmas Day (Figure 4.10), other major U.S. dailies did not print it so obviously, if at all. The *Chicago Tribune* placed the photograph, credited to Harrison Schmitt, on their Christmas Eve second page. Similarly, the *Los Angeles Times* pushed the image to the third page with no article and flanked by advertising and other news stories on Christmas Eve (Figure 4.11). And while the *Washington Post* regularly printed astronaut photographs through the days immediately after the landing and frequently about NASA topics, the *Whole Earth* image never appeared on their pages in late 1972.

Magazines of the day such as *Time*, *Life*, and *Newsweek* paid some attention to the Apollo 17 mission as well, but not to the same extent as early missions. Interestingly enough, the mission coincided with publication of the very last regular issue of *Life* magazine at the end of 1972. That final cover did not include any photographs, but did feature a large farewell story for Apollo, presumably because stories of human spaceflight and astronauts appeared so frequently and elegantly in *Life* issues throughout NASA's history. *Time* also included stories in their two issues immediately following the mission's end, but did not feature *Whole Earth* or Apollo 17 on a front cover. The December 11 *Life* issue that year had a caricature of Miami Dolphins head coach Don Shula on the cover, and a call out to a story about the end of Apollo in the cover's upper right corner. *Time* issues through the end of 1972 and into early January of 1973 focused cover stories on things ranging from skiing to Marlon Brando's latest film.[58] Similarly, the mission and photograph lost out to cover stories in *Newsweek*, which focused its mid-December through early January cover stories on the topics of crime, an Episcopal bishop, Henry Kissinger, and coping with depression. The crime issue of December 18, 1972, featured a similar story to that of *Time*'s December 11 issue, celebrating the end of Apollo, but additional issues featured still images taken from television coverage and not *Whole Earth* (December 26, 1972).[59]

The appearance of the *Whole Earth* image in deeper newspaper pages meant that different stories and photographs took precedence on front pages around the country. The ongoing Vietnam War continued to take up space, though coincident with Apollo 17 was word about a temporary halt to bombing Hanoi by U.S. forces for the Christmas holiday followed by intense bombing for multiple days that brought the North Vietnamese back to the table for peace talks. Major news also included daily updates on the medical condition of former President Harry Truman, who would pass away on December 26, discovery of plane crash victims who resorted to

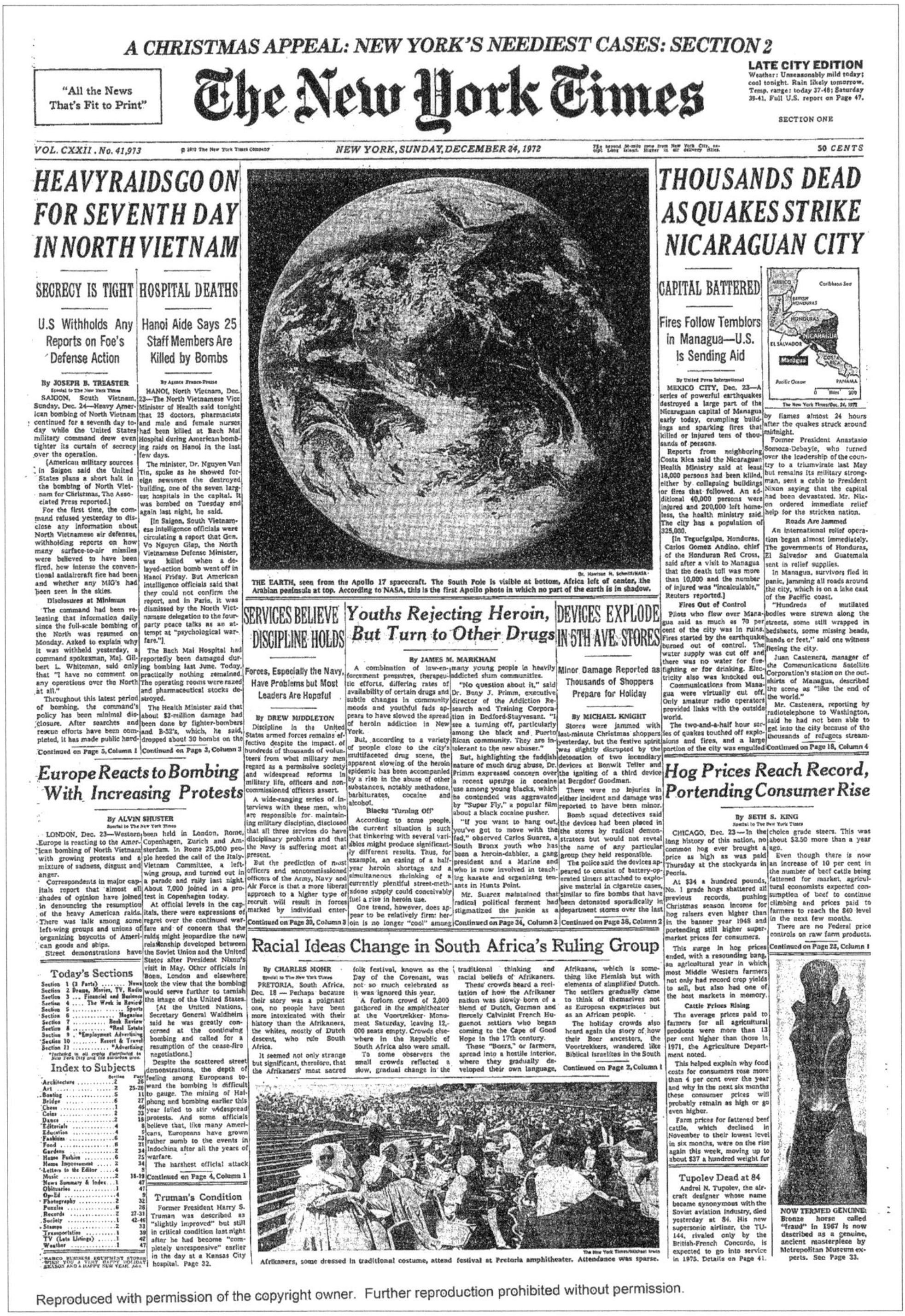

A CHRISTMAS APPEAL: NEW YORK'S NEEDIEST CASES: SECTION 2

"All the News That's Fit to Print"

The New York Times

LATE CITY EDITION

SECTION ONE

VOL. CXXII . No. 41,973 — NEW YORK, SUNDAY, DECEMBER 24, 1972 — 50 CENTS

HEAVY RAIDS GO ON FOR SEVENTH DAY IN NORTH VIETNAM

SECRECY IS TIGHT

U.S Withholds Any Reports on Foe's Defense Action

By JOSEPH B. TREASTER

HOSPITAL DEATHS

Hanoi Aide Says 25 Staff Members Are Killed by Bombs

THE EARTH, seen from the Apollo 17 spacecraft. The South Pole is visible at bottom, Africa left of center, the Arabian peninsula at top. According to NASA, this is the first Apollo photo in which no part of the earth is in shadow.

THOUSANDS DEAD AS QUAKES STRIKE NICARAGUAN CITY

CAPITAL BATTERED

Fires Follow Temblors in Managua—U.S. Is Sending Aid

SERVICES BELIEVE DISCIPLINE HOLDS

Forces, Especially the Navy, Have Problems but Most Leaders Are Hopeful

By DREW MIDDLETON

Youths Rejecting Heroin, But Turn to Other Drugs

By JAMES M. MARKHAM

DEVICES EXPLODE IN 5TH AVE. STORES

Minor Damage Reported as Thousands of Shoppers Prepare for Holiday

By MICHAEL KNIGHT

Europe Reacts to Bombing With Increasing Protests

By ALVIN SHUSTER

Hog Prices Reach Record, Portending Consumer Rise

By SETH S. KING

Racial Ideas Change in South Africa's Ruling Group

By CHARLES MOHR

Afrikaners, some dressed in traditional costume, attend festival at Pretoria amphitheater. Attendance was sparse.

Today's Sections

Index to Subjects

Truman's Condition

Tupolev Dead at 84

NOW TERMED GENUINE: Bronze horse called "fraud" in 1967 is now described as a genuine, ancient masterpiece by Metropolitan Museum experts. See Page 33.

Figure 4.9. *New York Times*, front page, December 24, 1972 (From the *New York Times*.

Guide to features

BRIDGE	95	LIVING	47-51
CLASS.	83-87	OBITUARIES	87
COMICS/XWD.	94, 95	SENIOR SET	30
DEATHS	88, 89	SPORTS	77-83
EDITORIALS	72	TV-RADIO	93
FINANCIAL	20	THEATERS	90-92

The Boston Globe

Ho, Ho . . . Hum
MONDAY — CLOUDY, 40
TUESDAY — CHANCE OF SNOW
SUNSET — 4:16
FULL REPORT — PAGE 22

Vol. 202, No. 178 © 1972, Globe Newspaper Co. · MONDAY MORNING, DECEMBER 25, 1972 · Telephone 288-8000 · 96 Pages—15c

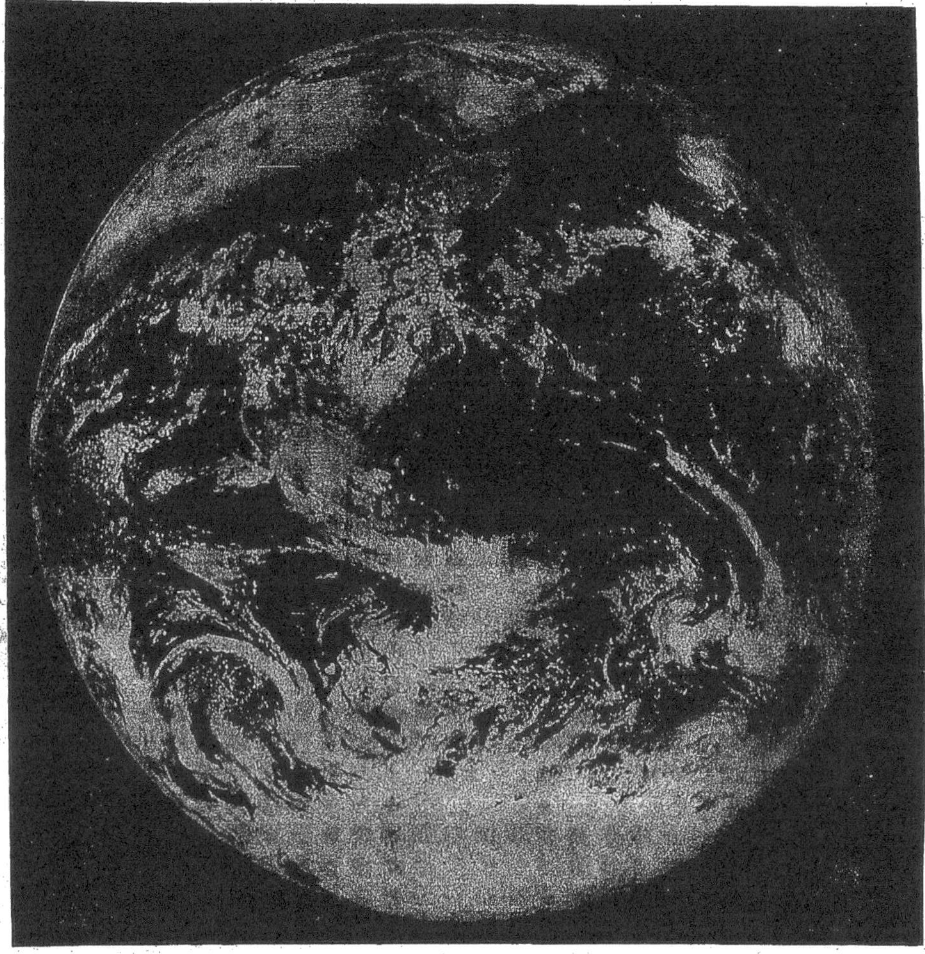

A Christmas editorial: 'It's the only Earth we have'

It's a weird and wondrous and awesome sight.

Captured on film by an Apollo 17 astronaut, the planet Earth hangs suspended in space like a giant tinsel Christmas ball, looking majestic and serene, solemn and stately, and at the same time so frail, so complex, so lonely, so confused—like the human race that inhabits it.

The Earth, and the life it sustains, is frail indeed—a speck of dust floating in the universe, infinitely tiny, breathlessly vulnerable. And so is mankind, so are we all. All equally mortal. All as close to death as a car collision on the way to the supermarket.

Like the Earth, spinning alone in the black void of space, we are all individually lonely, too. More so, perhaps, in this technical, computer-card age than ever before, our spirits crying out to each other for love and understanding and brotherhood and compassion. (Since everyone craves these things, secretly or openly, the question is why don't we give them to one another more freely and more sincerely, more of the time?)

The headlines on this Christmas morning, 1972, are full of havoc and heartbreak, some of man's own making and some beyond his control.

In Nicaragua, the death toll of the earthquake that ravaged that country's capital on Saturday continues to mount. Officials believe it may go as high as 12,000.

And on the other side of the world, in another little country, death and destruction of a different sort has escalated beyond redemption.

The world stands horrified, transfixed, by the savage, incredible renewal of bombing unleashed by President Nixon on North Vietnam. The "war" in Vietnam has been under way for so long it's as though everyone — including the participants—have forgotten what it's all about or why it began in the first place. Now, like a ghastly dream in a Kafka novel, it just goes on. And on. And on. And Americans can scarcely look the rest of the world in the eye, or themselves.

The President has declared a 24-hour bombing moratorium on Christmas day. Under the circumstances, however, the one-day truce is more revolting than reverent. The unprecedented mayham being wreaked on that people in the name of Peace is appalling beyond words, and it would be no worse on Christmas than it will be 24 hours later.

But not all the news of late has been bad. There is the account of the 16 rugby players who survived for more than two months on candy bars, roots and melted snow after their plane crashed in the Andes, 16 men holding stubbornly on to life and praying to God—whose mysterious intercession may have played a part in their miraculous discovery just two days before Christmas.

It is a story of human courage and grit and faith.

And of course there was the voyage of Apollo 17 itself, which produced the picture above. Like all the other manned space shots, this was a tribute to Man's ingenuity, imagination, energy and, one would hope, his humility as well.

As for the Earth—looking so turbulent, shrouded in clouds—here is a prayer for a Peaceful New Year. If the human spirit can conquer space, certainly we can find a way to live in harmony and compassion on our planet.

But this will only come about when we realize that life is sacred. And it is every bit as sacred in Quang Tri as it is in Duluth, Minn.

We've got to treat our Earth with love and care. It's the only one we have.

INSIDE:

Earthquake

The Nicaraguan government ordered a general evacuation of earthquake-stricken Managua, where the death toll was estimated to reach between 5000 and 12,000. President Nixon pledged assistance as relief supplies were flown in and Americans were airlifted to the states. Stories, photos, Pages 2, 3.

Bombing

For the first time in a week the skies over Hanoi were empty of bombers as air strikes on Vietnam were suspended for one day. The earlier loss of an 11th B52 was acknowledged. Story, Page 9.

Globe Santa

Donations from Globe Santa's friends have soared beyond the $300,000 mark. Story, Page 68.

Figure 4.10. *Boston Globe*, front page, December 25, 1972 (From the *Boston Globe*. © 1972 Boston Globe Media Partners. All rights reserved. Used under license).

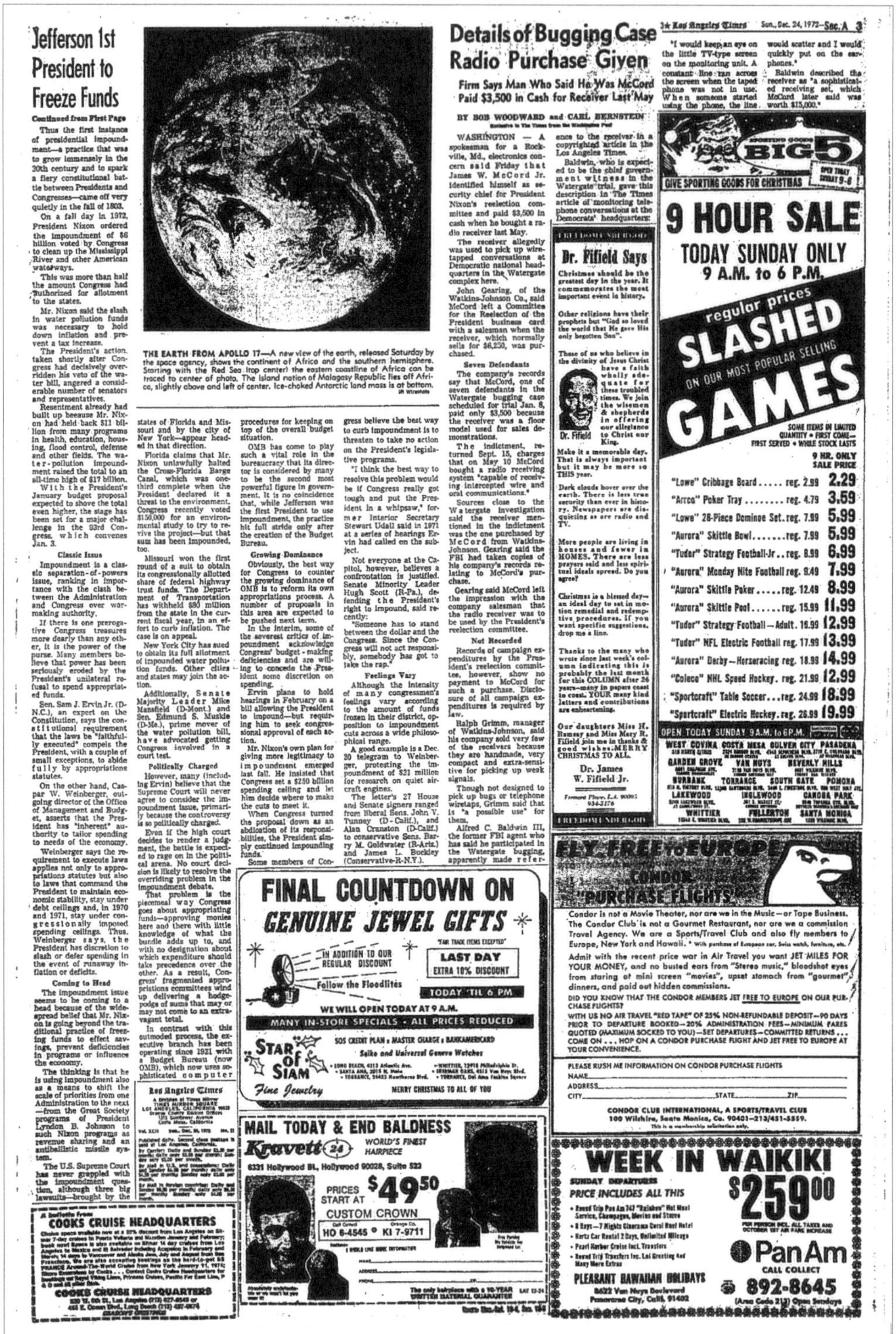

Jefferson 1st President to Freeze Funds

Details of Bugging Case Radio Purchase Given

Firm Says Man Who Said He Was McCord Paid $3,500 in Cash for Receiver Last May

BY BOB WOODWARD and CARL BERNSTEIN

THE EARTH FROM APOLLO 17—A new view of the earth, released Saturday by the space agency, shows the continent of Africa and the southern hemisphere. Starting with the Red Sea (top center) the eastern coastline of Africa can be traced to center of photo. The island nation of Malagasy Republic lies off Africa, slightly above and left of center. Ice-choked Antarctic land mass is at bottom.

Figure 4.11. *Los Angeles Times*, page 3, December 24, 1972 (Copyright © 1972, *Los Angeles Times*. Used with permission).

cannibalism in the Andes Mountains, and a 6.2 magnitude earthquake that struck Managua, Philippines. After five previously successful missions to the lunar surface and one near-disaster, perhaps NASA images and stories no longer elicited the same emotional reactions they did during the triumphal moments of Apollo missions 8 and 11: the shine of human spaceflight had worn off. Therefore, while research material for scientists and engineers skyrocketed following the excitement of the early lunar missions, newspapers began treating NASA human flights as a worthy pursuit for stories only when passing significant landmarks. Since publishers especially considered events only in the short term, looking at the context of the image provides a sense of what else meant something to people at the time. Scholarship that considers this period broadly tends to ignore space exploration as part of the social, economic, and political landscape despite its ubiquity in mass media.

ONGOING REACTIONS

Despite a relatively rapid drop in front page attention paid to human spaceflight between Apollo missions 8 and 17, astronaut-captured photographs made a lasting impression on public memory and represent some of most instantly recognizable and beloved images of the twentieth century. While novelists, artists, and cartoonists imagined a visual future for humans leading up to and through the beginning of Space Age, astronauts brought home images for us to participate virtually and visually in their journeys. Ideas about space travel, our world, and the Moon took shape through the lenses of astronauts, and then on the pages of newspapers, magazines, and publications that reached around the world to spread positive impressions of NASA's exploration efforts. The penetration of these images into our collective memory not only serves as a testament to the beauty of our environment, but also reminds us of the turn toward an ever-increasing visual culture during the 1960s.

Rarely, if at all, have historians included the details or even general facts about the space program in survey histories of the 1960s. Most major works barely make note in accompanying timelines of the seminal moments of say John Glenn's Project Mercury flight or Neil Armstrong's steps on the lunar surface during Apollo 11.[60] Issues of technological development, political wrangling, or the larger cultural significance of spaceflight are left to specialist historians to examine, and then relate to the larger historical narrative. So despite the tremendous reach of astronaut photography through publications and reuse in other media, and its ongoing role as a touchstone for public memory of the early NASA human spaceflight program, it never received more than a passing appearance as an illustration in survey scholarship on the Cold War.

Worth consideration is how the use of color or black-and-white film altered conceptions of features seen in astronaut images. Media formats such as magazines and indexical catalogs offered some exposure to full color photography, but the public most often saw still photographs by astronauts published in black-and-white as part of newspaper mission coverage. Magazine subscribers for *Time*, *Life*, and *Newsweek* benefited from regular large, multipage color spreads of selected images and narratives about the missions.[61] Special editions of these magazines featuring even more photographs were certainly popular with nonsubscribers at newsstands and in grocery stores as mementos of the events. When it came to the Moon, however, photographing in black-and-white had more to do with scientific needs than engineering or public uses.[62] Printed black-and-white photographs and poorly color corrected stills may have unintentionally fostered misconceptions about the actual color of the Moon. In different recordings, lunar astronauts described the color of the Moon as varying from tan to a dark gray depending on Sun angles.[63] The composition of dark lunar dust and rocks, reflective of the Sun's light, tends to contain more glass particles containing iron, giving lunar regolith a brown or rust-colored appearance, and making it magnetic.[64] Unless casual viewers read additional materials, though, the impression left of the Moon was a very bland, bleak, and colorless one indeed based on cursory viewings of photographs.

How then would such an impression of the Moon, plainly seen in prints in newspapers and from the early black-and-white television cameras used during the lunar missions, negatively influence what appeared to be a very popular and exciting national initiative? While even the ability to use television and still cameras in such harsh and cramped places was a technological triumph, their products did little to sell space as vibrant and welcoming. In truth, the Earth came off as beautiful and fragile and the Moon dull and desolate. Nothing about astronaut photographs could change how the mood of American society would influence interpretations of the images, and come to understand them within the context of events of the time. These images represent a brief, symbolic, shining moment in time when humanity achieved something spectacular. Neither the journey nor the destination was attractive enough to keep public attention focused on the Moon as more than a place we once went.[65] The ultimate reality of Apollo could be that the Earth and places that lay beyond our Moon, seen in spectacular images taken by robotic explorers and space telescopes, would be the real objects of attention for NASA.

Several instances, however, are worthy of note where astronaut photographs inspired the creation of artwork, public art displays, and other public presentations, although these impressions were not commonplace or accessible to very many at the time. Syndicated opinion pieces and books like Norman Mailer's *Of a Fire on the Moon* were widely available in addition to news broadcasts and other media products that directed attention at perceived problems with flying people to the Moon or questioned its value in comparison to terrestrial concerns.[66] While these pieces largely reflect broadly upon actions taken by NASA, there is significant evidence of contemporary recognition of how astronaut photography influenced ideas about the Earth, the Moon, and ourselves. Anne Collins Goodyear reflected on NASA's adoption of an art program early in its life in her dissertation and resulting book chapter, "NASA and the Political Economy of Art, 1962–1974."[67] Artists frequently drew inspiration from Apollo photographs, especially as seen in works by Angela Manno (*Home*, Figure 4.12), Derman Uzunoglu (*Earth Rise*, Figure 4.13), and Robert Shore (*Lunar Confrontation*, Figure 4.14), which all now reside in the collection of the Smithsonian's National Air and Space Museum. Even before Apollo's end, scholars and artists recognized this connection and the expectation that visuals from the space program would continue to inspire artists for years to come.[68]

In addition to those standalone pieces, some photographs provided inspiration for larger artistic works in public spaces. In the 1970s, Drexel University in Philadelphia, PA, installed a mosaic representation of an Apollo 11 *Earthrise* image at the entrance to their Newman Center building, which was removed when it became the Marks Intercultural Center in 2010 (Figure 4.15).[69] Robert McCall, who painted conceptual art for the movie *2001*, made a career out of imagining and reimaging the known and unknown in space. While it is not a literal interpretation of an astronaut photograph, McCall's mural *The Space Mural—A Cosmic View*, painted in place by the artist for the opening of the National Air and Space Museum in July 1976, represents visual elements of the Apollo lunar experience and space sciences (Figure 4.16). Such elaborate works are rare and do not typically exist outside museums and NASA facilities, so impressions made on viewers occur only in environments already attracting a self-selected audience of those interested in spaceflight. But the scarcity of large public displays such as these shows, to some

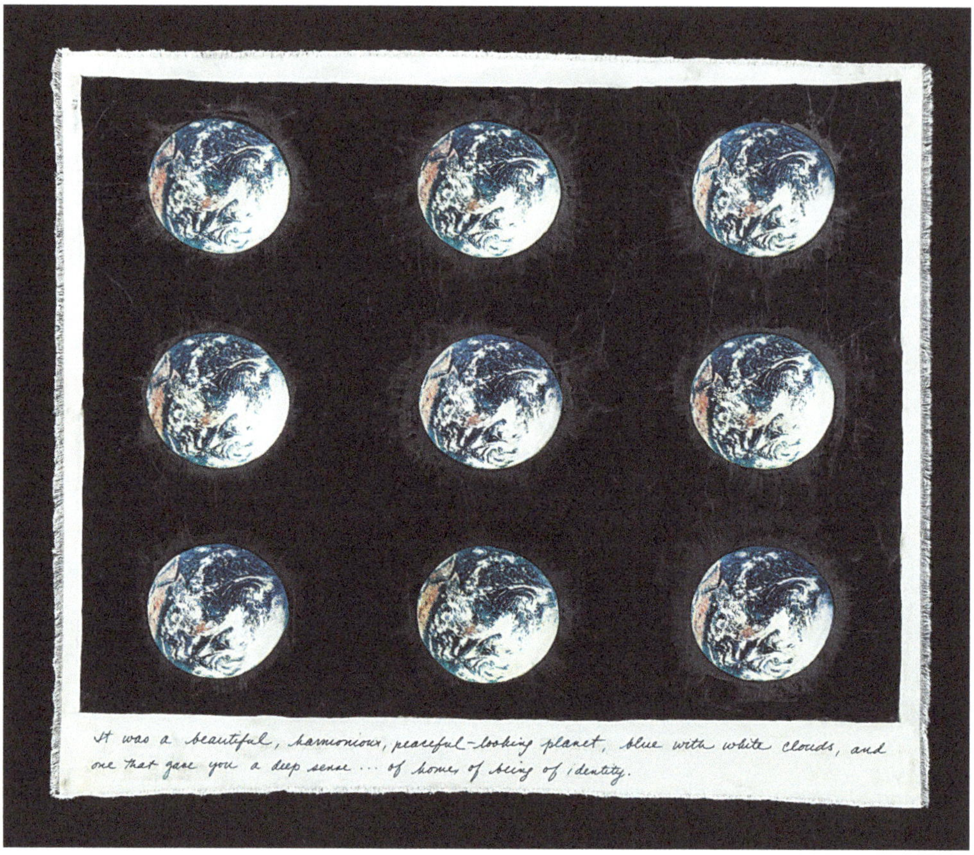

Figure 4.12. *Home* by Angela Manno, batik and color xerography, 1985, NASM Cat. #A20030012000 (Angela Manno).

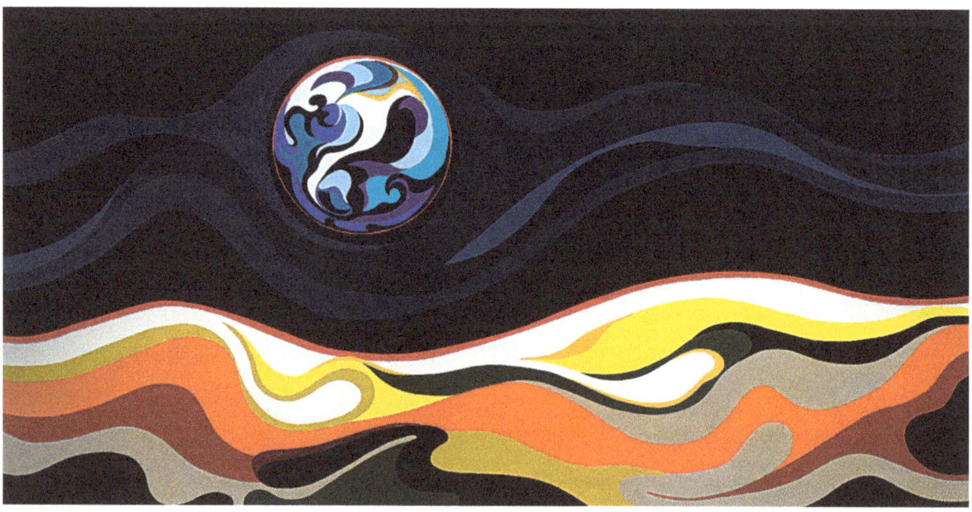

Figure 4.13. *Earth Rise* by Derman Uzunoglu, acrylic on canvas, 1974, NASM Cat. #A19760076000 (Smithsonian National Air and Space Museum)

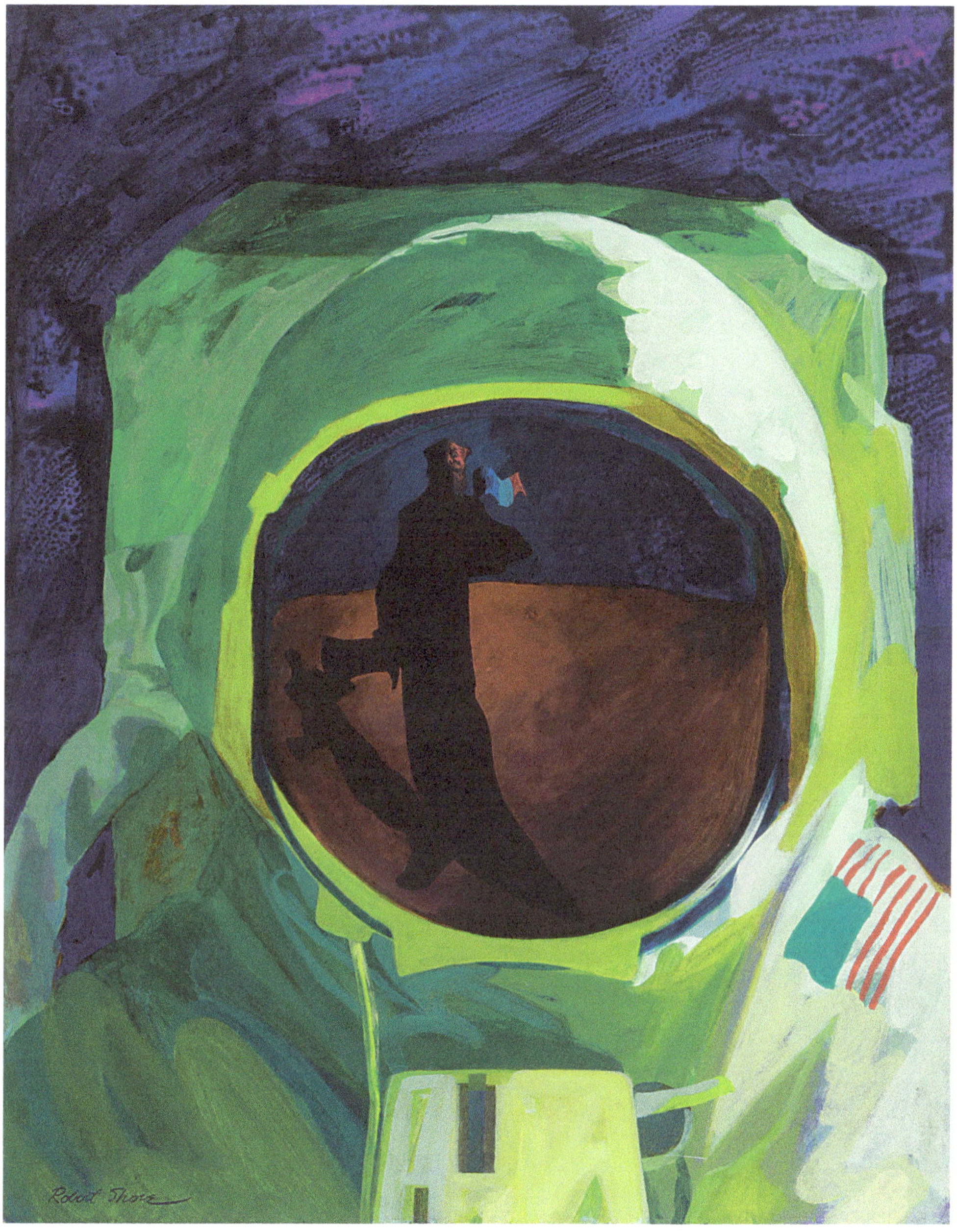

Figure 4.14. *Lunar Confrontation* by Robert Shore, oil on masonite panel, 1970, NASM Cat. #A19760332000 (Smithsonian National Air and Space Museum).

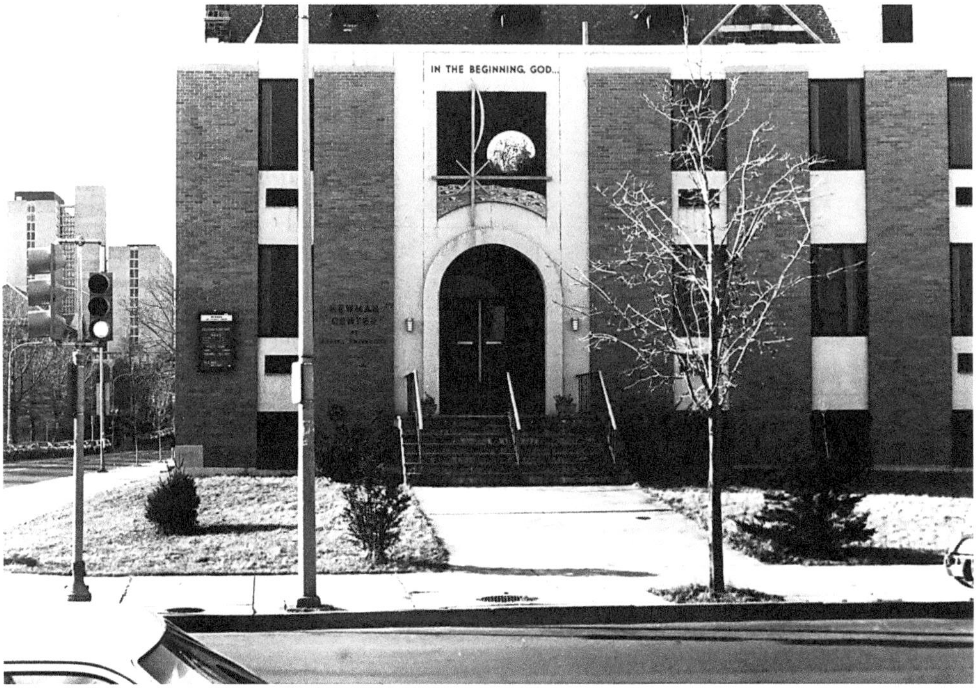

Figure 4.15. Mosaic representation of *Earthrise* image, Newman Center, Drexel University, removed 2010 (Drexel University Archives; I. George Bilyk).

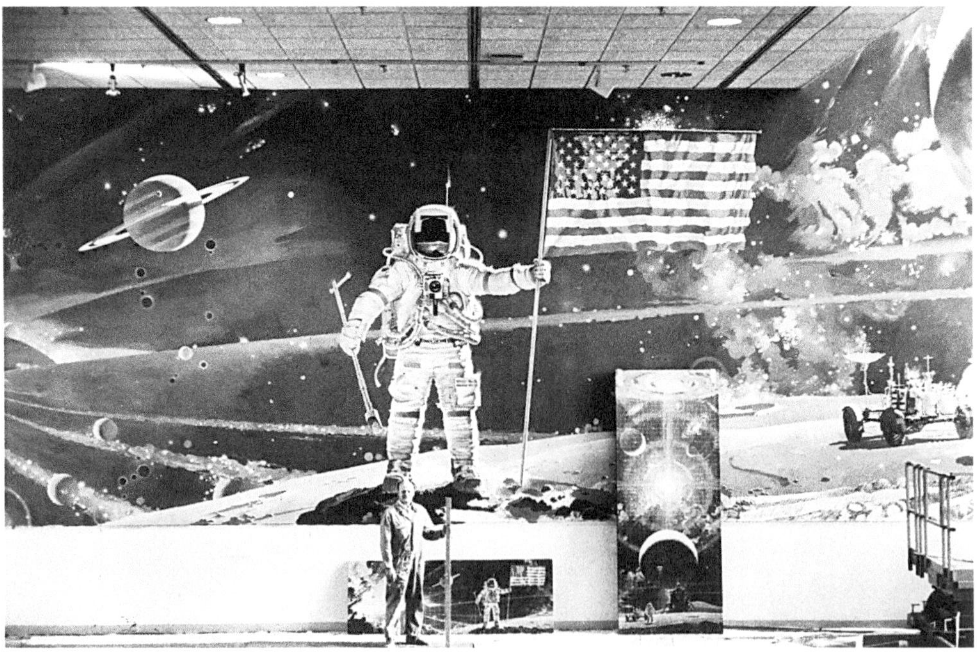

Figure 4.16. *The Space Mural—A Cosmic View*, painted by Robert McCall, 1976, NASM Catalog Cat. #A19780181000 (Smithsonian National Air and Space Museum).

degree perhaps, that while space artwork and photography have tremendous appeal to a public audience in small personalized settings, there may not be a place for human spaceflight in large public settings not directly related to spaceflight.

While some were inspired to the point of creating artwork, others saw those photographs as information and messages for motivating a movement. Members of the environmental movement rejected the frontier and imperial overtones of the images in favor of a more thoughtful approach to understanding the Moon-Earth relationship. Recent historical scholarship by Robert Poole and Neil Maher on the meaning of Apollo photographs to intellectual and environmentalist conceptions of Earth contribute to our understanding of images as tools for shaping philosophical and political beliefs. These images, and what it took to acquire them, affected American culture in profound ways, both answering and provoking questions about our place in the universe and our care of "spaceship Earth."[70] Maher makes a particularly salient case for where the NASA narrative of this period overlapped with other political and cultural movements, and uses *Whole Earth* as a case study of the often oversimplified understanding of the relationship between the environmental movement and the actions of the space agency.[71] What lies beyond the horizon in *Earthrise* is our home planet, and seeing this sight for the first time through astronaut eyes gave social commentators like Anne Morrow Lindbergh reason to ponder the fragility of our floating blue orb. For geographers as well, seeing the Earth from great distances gave them an avenue for conceptualizing how we understand the Earth in physical terms. Geographer Denis Cosgrove observed that "They [Apollo photographs] have been enormously significant however in altering the shape of contemporary geographical imagination."[72] Conceiving anew our view of the physical Earth coincides with the reconsideration given to the well-being of the planet.

My contention regarding the space agency's use of photography generally aligns with Kim McQuaid's statement that "From NASA's formative years, it had consistently mishandled opportunities to increase its political support by providing practical and understandable Earthly services to citizens and taxpayers."[73] By missing an opportunity to work collaboratively or at least transparently with environmentalists and taxpayers with Earth photography, whether from satellites or on human missions, NASA moved away consciously and publicly from ideas about Earth in favor of continued exploration outward. Other evidence also suggests that top administrators in NASA, Hugh Dryden in this case, rejected vigorously the idea of including Earth photography on human spaceflight missions. Despite the obvious political implications of the use of *Earthrise* as an image of a triumphant American project, this moment also signaled NASA's passivity toward the opportunity to fuel Earth research during the formative years of climatological studies. While astronauts and scientists continued working on Earth-focused imaging, administrators made little attempt to make a firm statement about such observations, perhaps to avoid the politics. That all changed, however, when exploration goals set for NASA by Presidents Eisenhower and Kennedy turned more abstract, this lead to projects aimed to use up remaining Apollo hardware (Skylab and the Apollo Soyuz Test Project). The seeming extravagance of lunar voyages gave a pragmatic appeal to supposedly more affordable orbital missions using a reusable space truck that turned attention back toward a more attainable and understandable Earth environment.

Understanding meaning in the case of *Earthrise* and other astronaut photographs involves acknowledging the multiplicity of meanings possible when NASA released images to the public with little context and supporting information. When picked up by the press, social commentators, and others, an image's popularity skyrocketed, and people found their own meanings within the codes embedded in the visual. Public reception was and continues to be very positive for images that astronauts captured, and which are perhaps some of the greatest products of human spaceflight. When asked what he thought his most significant contribution to the space program was, astronaut and *Earthrise* photographer Bill Anders modestly admitted, "Maybe taking that picture which had a lot of ecological and philosophical impact at the time."[74] Even the photographer himself understood the historical significance and long-term influence of our scientific and cultural perspectives on the Apollo program, perhaps better than NASA itself.

EPILOGUE

Continuing Resonance

> *"People who go to these frontiers want to share the experience, they want to record data to record the knowledge from the frontier, and photography is a great medium in which to do this."*
>
> —*Dr. Donald Pettit, astronaut, photographer, and chemical engineer, Luminance Conference 2012*

In the fall of 2008, I traveled to the Kennedy Space Center with colleagues on a team creating a new exhibition, *Moving Beyond Earth*. Our topic, making low-earth orbit a permanent place to live and work starting with the Space Shuttle, made the space center an excellent reference point for things never before examined in our museum. The diverse spaceflight workforce, decision-making in spacecraft design, and reusable technology were on our minds during a guided tour of the processing building for the orbiter *Discovery*, a ride up the launch pad elevator to the white room and the open hatch of *Atlantis*, and a walk around the ISS equipment facility. These opportunities gave us unforgettable first-person views of all the hardware it takes to keep the United States in space.

During our tour of the ISS facility, we saw payloads being prepared for launch, and the huge containers used to transport and load them into the Space Shuttle. One of those payloads caught my eye, something called the cupola (Figure Epi. 1). I had heard of it not only for its *Star Wars*-like configuration of windows, but also because this was to be an almost entirely experiential space, meant to give astronauts unfettered views of Earth with one round central and six trapezoidal surrounding windows.[1] But at the time, it was unclear if the unit would ever go into space at all. The lead exhibition curator, Dr. Valerie Neal, commented on my excitement over photographs of the cupola. She told our group that expectations inside NASA at the time were that in the post-*Columbia* tragedy climate of caution, payloads going to the ISS were streamlined to only those necessary for science and servicing. We were all sad to hear that this amazing ISS attachment might never move past the spot where we saw it in Florida.

Learning about this decision reinforced my sense of NASA's uses of and expectations for photography. What mattered was scientific research and not the potential for visual rhetoric to create emotional experiences related to seeing Earth and space. The post-Apollo era of human spaceflight has been marked by an attitude of "what have you done for me lately" in terms of payloads released, science results, and real-world benefits of spaceflight. Construction of the ISS fell into that same category, with only the disruption to assembly caused by the loss of *Columbia* on a science mission in 2003. Nevertheless, despite over four years of delays following the completion of the cupola in 2005 by Italian subcontractors, NASA confirmed that the cupola would

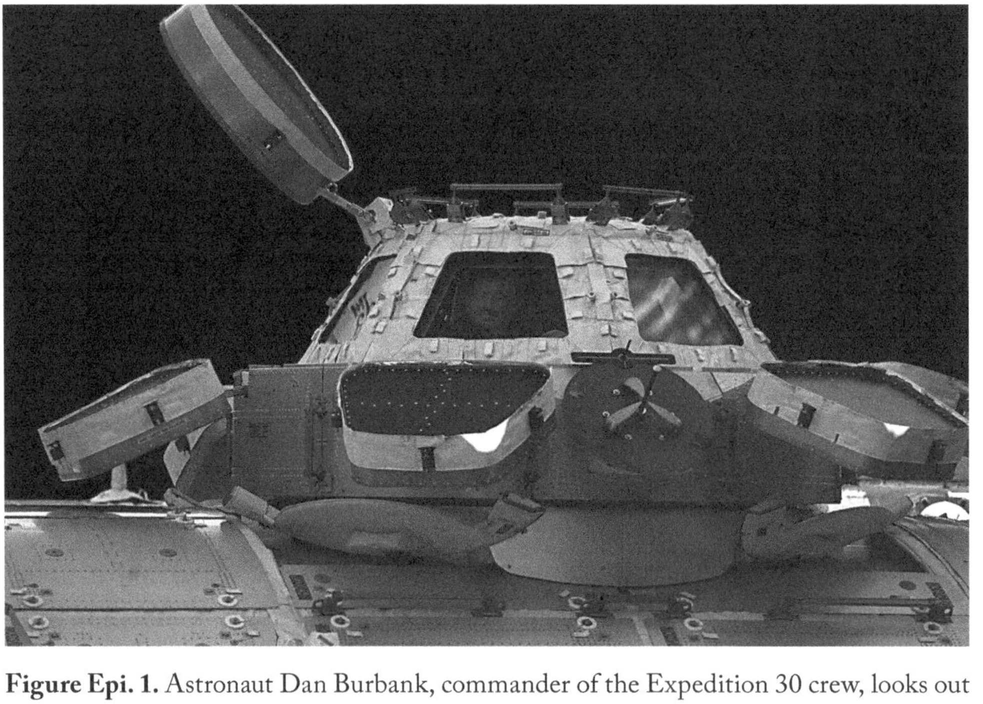

Figure Epi. 1. Astronaut Dan Burbank, commander of the Expedition 30 crew, looks out the cupola windows, April 21, 2012, ISS030-E-270467 (NASA).

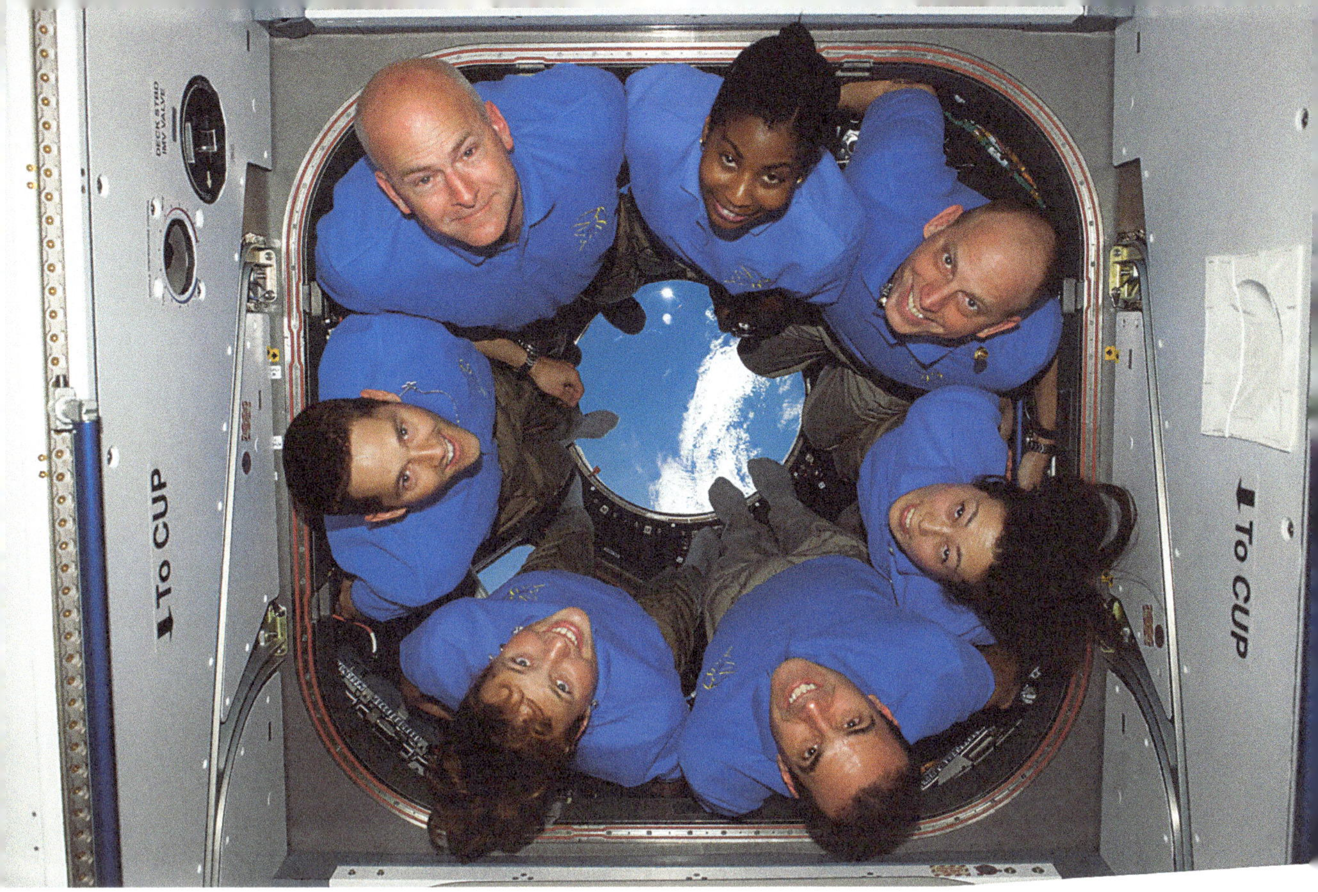

Figure Epi. 2. Each STS-131 crewmember "stands" in one of the angled cupola windows, April 14, 2010, S131-E-010051 (NASA).

indeed fly to the ISS. Once installed during STS-130 in early 2010, it quickly became a gathering place for astronauts and frequent location for crew portraits (Figure Epi. 2). It also gave astronauts interested in photography a platform for documenting Earth, natural phenomena, and their work from the unique vantage point of 350 kilometers (220 miles) altitude. The availability of such a perspective brings the story of handheld astronaut photography full circle from the tentative experimentation on the flight of John Glenn in 1962 to the work of the latest ISS crews today.

AFTER APOLLO

From the first orbital flight through trips to the Moon to orbiting Earth daily, astronaut photography fundamentally shaped how we see and understand space, spaceflight technology, and astronaut life. The end of the Apollo program was far from the end of when its images could affect people. The volume of photographic work on Apollo 17 signified an exponential increase of still frames returned. Afterward, astronaut photographs turned exclusively to near-Earth photography with orbital stays starting with the Skylab Orbital Workshop. The vantage point of orbit, be it from Skylab, the Space Shuttle, or ISS, did not eliminate opportunities for astronomical or geological photography, but refocused them on the Sun and Earth. The emergence of remote sensing, weather, and other satellites and telescopes removed the vast majority of mapping and scientific photography from the bailiwick of astronauts, allowing them to train their cameras on more targets of opportunity. Thanks to additional storage space on bigger spacecraft, astronauts were virtually unconstrained by the amount of film available. After digital cameras came into use in the late 1990s and exclusively with the operational phase of the ISS in the early 2000s, they were restrained only by the amount of electronic storage space or download capacity from the station to Earth.

Recalling the heroic age of human spaceflight today falls largely on the whims of the media, as seen in celebrations of the Apollo 11 fiftieth anniversary in 2019. But others affected by the images and experiences of that period make their own connections to the visual culture through art, presentations, and verbal recollections. For his part, Alan Bean was the only astronaut of the early program who made a living off what he saw during his time in space. While he took art lessons before his Apollo 12 flight, he spent time after his astronaut retirement creating real and imaginative views of that period with oils on canvas (always including a trace of what he claimed to be a little Moon dust from a patch he wore on his lunar spacesuit). His catalog of original paintings numbers in the hundreds, and lithographic reproductions are available widely online and in specialty space-related stores. The printed catalog of an exhibition of his works, *Painting Apollo*, shows his continued interest in commemorating his experiences and those of his fellow astronauts, and providing the public with a first-person interpretation of the spaceflight experience. Though only sometimes based strictly on photographs taken during missions, his work played on popular ideas of the heroic but never questioned the value of the early human spaceflight program.

A more diverse and experienced set of astronauts emerged after Bean's time. Astronauts rode on spacecraft capable of carrying up to eight passengers at a time. The Space Shuttle, which NASA deemed a utility truck for launching satellites and other payloads into orbit in its early years, became an intermittent orbital station for astronaut photography. NASA extended imaging duties to include shooting films that benefited commercial outlets, using precious training and on-orbit time for such ventures as movies using IMAX equipment.[2] Not only did astronauts continue as still photographer surrogates, but they became trained proxies for cinematographers. The limited amount of 70mm IMAX film carried to space in storage lockers meant that directors and cinematographers planned virtually every second of film use. Toni Myers, writer or director of most of the NASA series of IMAX movies, commented that the astronauts were so well trained and captured such stunning footage that barely a frame of what was captured over 24 flights went unused for the six movies they made with NASA cooperation.[3] Unlike the days of NASA leadership having to beg astronauts to carry out live television broadcasts, shuttle astronauts willingly participated in creating stunning visual representations for gigantic screens that put people closer to the feeling of spaceflight than ever before.

Just like astronauts during the first decade of human spaceflight at NASA, some shuttle and space station crewmembers enjoyed and focused on photographic tasks more than others. Neurologist Roberta Bondar, the first female Canadian astronaut, flew to space in 1992 on the Space Shuttle and parlayed her passion for photography into a post-astronaut career in photography.[4] Space tourist Richard Garriott undertook a massive photographic project during his eleven-day stay on the ISS. Garriott's father, Skylab and shuttle astronaut Owen Garriott, photographed a series of locations on Earth during his fifty-nine days of the second Skylab mission. His son attempted to capture the exact same locations as part of the Windows on Earth project to show the dramatic change in its surface during the intervening thirty-five years.[5] These examples show how an interest in photography could collect useful information, but not necessarily as an artistic endeavor.

Only toward the end of the Space Shuttle program did any obvious plans arise to have astronauts spend more than a passing moment considering the same issues as professional art photographers. A project between art photographer Michael Soluri and the crew of the last Hubble servicing mission, STS-125, brought art and photography very close to merging.[6] Soluri proposed photographing their training, shooting a series of portraits of the crew and training staff in the common and uncommon settings around NASA facilities. At the request of mission commander Scott Altman, he also taught the crew about considering image composition, to see more than just information but a representation of the experience. The results were spectacular, with a self-portrait by astrophysicist and amateur photographer John

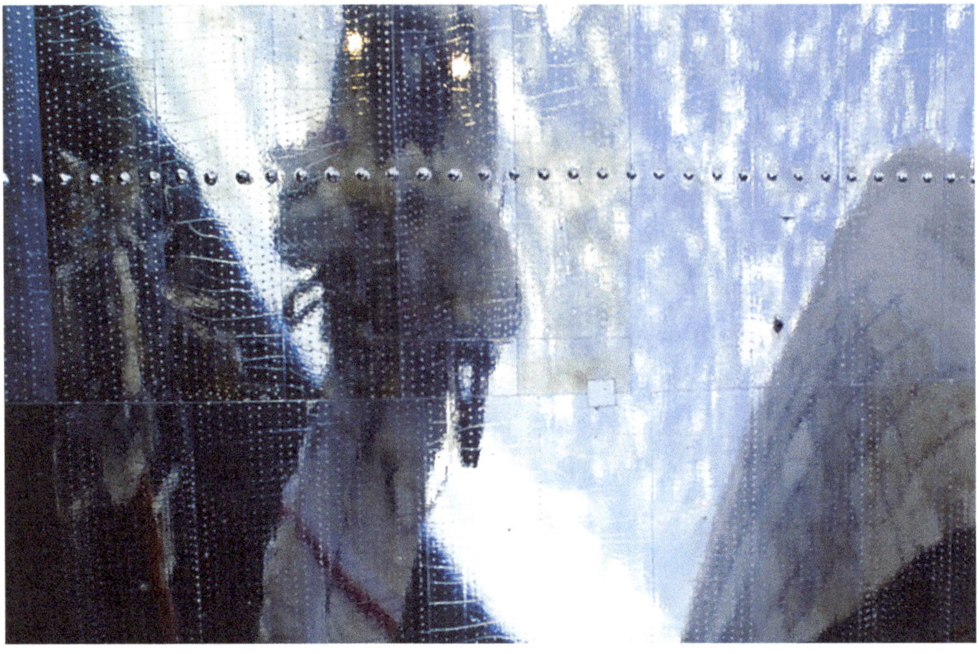

Figure Epi. 3. Astronaut John Grunsfeld photographed his reflection in the side of the Hubble Space Telescope, May 18, 2009, S125-E-010077 (NASA).

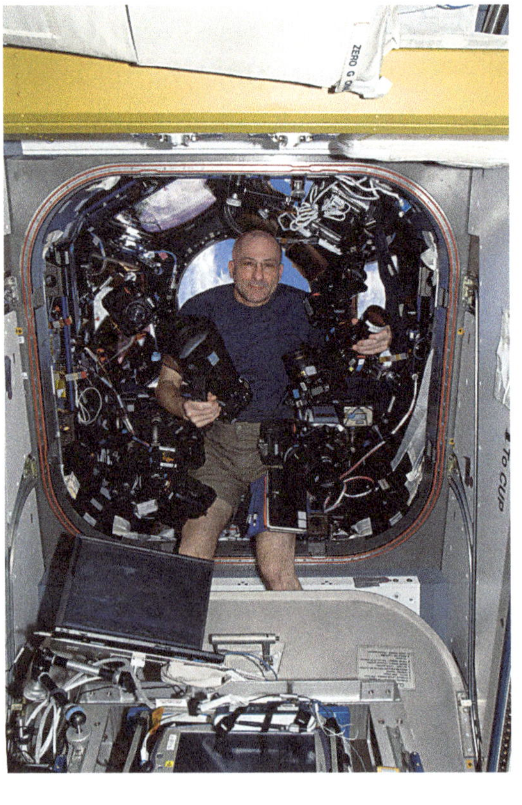

Figure Epi. 4. Astronaut Donald Pettit positioned inside the cupola and surrounded by camera equipment, June 10, 2012, ISS031-E-112469 (NASA).

Grunsfeld showing the great potential for astronauts to develop as artists with the camera (Figure Epi. 3). That training in photographic techniques resulted in their mission presenting a perspective not typically taken to space by astronauts.

An astronaut with an artist's eye, technical expertise, and uncommon ingenuity broke down the wall between being a proxy to being a legitimate photographer. Two-time ISS resident Donald Pettit capitalized on his time and the prime Earth viewing location of the cupola more than anyone has so far (Figure Epi. 4). For his second mission, Expedition 30/31 in 2012, he used a mounting system inside the cupola for up to seven cameras to shoot simultaneous still photographs to create a stream of images to run like a high-resolution video of the aurora and weather patterns.[7] Pettit lectures about his experience and photography, describing the cupola as one of the most unique and meaningful facilities on the ISS.[8] Much like astronaut-artist Alan Bean, Pettit took an active role in sharing life in space through an astronaut's eyes thanks to camera equipment.

The visual products of our proxies in space, going back to John Glenn, continue to resonate with the same audiences NASA wanted to reach when the images first returned from spaceflights. Perhaps in this age of rapid digital reproduction, the most iconic of astronaut photographs appear in hundreds if not thousands of online presentations. Those who take up a particular space-related cause, just like Stewart Brand and his push for a whole Earth image, regularly use astronaut photographs to illustrate their points. NASA feeds this interest with their own websites dedicated to astronaut Earth photographs.[9] *An Inconvenient Truth* featured interpretations of *Earthrise* and *Whole Earth*, images that became touchstones of Apollo in our collective memory as demonstrative of the fragile and pure Earth photographed by the heroes of Al Gore's and so many of his generation.

Gore was far from the only public figure of his generation to look back fondly on early human spaceflight for inspiration and meaning. Dr. Neil deGrasse Tyson, director of the Hayden Planetarium at the American Museum of Natural History in New York City, speaks about the influence of *Earthrise* on his own life, the trajectory of spacefaring, and our culture. About halfway through his keynote address from the twenty-eighth annual National Space Symposium in 2012, he commented on the Apollo 8 mission, its place within the story of 1968, and the image that he says changed forever the way people look at Earth. Tyson stated: "There was Earth, seen not as the mapmaker would have you identify it. No, the countries were not color coded with boundaries. It was seen as nature intended it to be viewed. Oceans, land, clouds. We went to the Moon, and we discovered Earth. I claim we discovered Earth for the first time."[10] His lifelong passion for spaceflight, astronomy, and science fiction resonates among the space enthusiast community, but his personable style allows his words to appeal broadly. Tyson's commentary often goes viral, and this small segment about Apollo 8 became the core of a campaign by NASA and space enthusiasts to reinvigorate support for spaceflight.[11] As was true in the 1960s, however, concerns over economic stability, national security, and social welfare seem to have drowned out voices in favor of a strong spaceflight program.

Where NASA finds the most success now with images are not those released from astronaut handheld cameras, but rather from those remotely operated from Earth. The popularity of the Hubble Space Telescope and a series of rovers on Mars indicate a revived enthusiasm for space exploration, but perhaps not one that involves the same risks as using humans as the source of image collection. Extending our vision to other planets, solar systems, and galaxies continues a legacy of imaging that goes back to NASA's beginning, but with an abstract person (or set of people in these cases) behind the controls of the process, we cannot imagine ourselves in the shoes of a photographer. An analysis comparing astronaut photography to the incredible popularity of Hubble and rover photographs would be informative, especially regarding the cultural attachment created between the public and these robotic proxy observers. The subject, however, requires additional research not part of the line of inquiry laid out here.

CONCLUDING THOUGHTS

We remember the Space Race of the 1960s as a series of dramatic successes and some failures, events seen around the world on television and the front pages of newspapers and magazines, and duplicated on websites and publications today as reminders of a heroic age of spaceflight. While millions watched from afar and hundreds of thousands took part in making those moments possible, only a handful of people rode rockets into space and served as proxies for the professionals and public left behind. What astronauts returned to Earth as representations of their experience, all of the textual and visual information on which we base our understanding of life off this planet, continues to serve American culture as a source of pride and wonder at those achievements.

Looking to exploration as a source of understanding in our culture goes back to the earliest recorded ventures to unknown parts of the world. With the availability of cameras and photographers to document the landscape, expedition leaders gave supporters evidence of development potential and the public truthful representations of places they could not travel to themselves. For each of these projects, common themes bound the visual depictions and similar characteristics carried through a century of human movement into unknown places. Those people selected to photograph earlier expeditions on Earth were trained professionals, well prepared with their own ideas about what audiences should see of extreme environments. Astronauts came from the ranks of test pilots, and later scientists, rarely bringing any experience with cameras with them to NASA. As our surrogates in space, only the best training and equipment would suffice for their trips to orbit and the Moon. The images astronauts returned more than satisfied the needs of most audiences. After passing through successive filters of managers, public affairs specialists, and the media, the photographs provided the public and our collective memory with icons of the Space Age. Those photographs became part of our visual lexicon of spaceflight, appearing in the most widely circulated publications of the time, and thousands of publications and websites since then. Although space images have been ignored by historians of the 1960s, the instant recognition of the stories contained in these iconic images compelled this study, an attempt to contextualize astronaut photography in a wider historical narrative.

When looking at photographs taken with a Hasselblad camera by the first handful of humans to fly into space, one might notice parts of Earth or the Moon never noticed before with incredible clarity. Most of those images, in their original run and in recent incarnations when digitized for mass consumption, rarely saw the light of day beyond more than a few hundred scientists' eyes. This means the bulk of the photographic work of astronauts meant relatively little as a contribution to collective memory of the early human spaceflight program. With images as their most significant contribution to the public understanding of what space feels and looks like, astronauts willingly took up the challenge of fulfilling the needs of virtually every person on the planet, serving as our eyes in space. Considering astronauts did this with the world observing their every move, their photographs cannot be considered as any less than sublime.

Notes

INTRODUCTION: INTERPRETING ASTRONAUT PHOTOGRAPHY

1. "The Heron and the Astronaut," Lindbergh's essay on Apollo 8, appeared in *Life* magazine on February 29, 1969. Her publisher combined it with "Immersion in Life" from *Life*'s October 21, 1966, issue into a single bound volume: Anne Morrow Lindbergh, *Earth Shine* (New York: Harcourt, Brace & World, 1969), 45.
2. Anders captured this color image as the spacecraft traveled around the Moon near its equator. Its orientation reflects his position in the spacecraft relative to the Moon. A black-and-white photograph taken moments before by Frank Borman shows the Moon oriented below the spacecraft. Anders stated frequently that the Moon was on the left in his view. Published versions of the image show either the Moon to the right or the scene as a landscape. None are right or wrong since there is no up or down in space.
3. Michael F. Robinson, "Lessons from the Last Frontier," Keynote Address at "1961/1981: Key Moments in Human Spaceflight" symposium, April 26, 2011, Washington, D.C., http://www.ustream.tv/recorded/14289321 (accessed June 29, 2012); Anders Houltz, "Selling the Space Camera., Society for the History of Technology Annual Conference, October 17–21, 2007, Washington, D.C.
4. Robert K. Poole, *Earthrise: How Man First Saw the Earth* (New Haven: Yale University Press, 2008); Benjamin Lazier, "Earthrise: or, The Globalization of the World Picture," *The American Historical Review* 116, no. 3 (June 2011): 602–30.
5. Poole, *Earthrise*; Dwight Steven-Boniecki, *Live TV from the Moon* (Burlington, Ont.: Apogee Books, 2010).
6. Elizabeth Kessler, *Picturing the Cosmos: Hubble Space Telescope Images and the Astronomical Sublime* (Minneapolis: University of Minnesota Press, 2012).
7. Steven-Boniecki, *Live TV from the Moon*; Heinrich Freytag, *The Hasselblad Way; the Hasselblad Photographer's Companion* (London: Focal Press, 1973); Alain Lazzarini, *Hasselblad and the Moon* (Granville, France: Self-Published, 2011).
8. Philip Scranton, "None-Too-Porous Boundaries: Labor History and the History of Technology," *Technology and Culture* 29, no. 4 (October 1988): 722–43; David A. Mindell, *Digital Apollo: Human and Machine in Spaceflight* (Cambridge: MIT Press, 2008).

9. Ruth Schwartz Cowan, "The Consumption Junction: A Proposal for Research Strategies in the Sociology of Technology," in *The Social Construction of Technological Systems: New Directions in the Sociology and History of Technology*, eds. Wiebe E. Bijker et al. (Cambridge: MIT Press, 2012), 253–72.
10. David Nye, *American Technological Sublime* (Cambridge: MIT Press, 1994).
11. Roland Barthes, "Rhetoric of the Image," in *The Visual Culture Reader*, ed. Nicholas Mirzoeff (London: Routledge, 2002), 136.
12. Sigmund Freud, *The Standard Edition of the Complete Psychological Works of Sigmund Freud*. Vol. III. (London: Hogarth Press, 1953), 303–22.
13. Siegfried Kracauer, "Photography," trans. Thomas Y. Levin, *Critical Inquiry* 19, no. 3. (Spring 1993): 421–36.
14. Michel Foucault, "Panopticism," in *Visual Culture: The Reader*, eds. Jessica Evans and Stuart Hall (London: Sage in association with the Open University, 1999), 61–71.
15. Terry Barrett, "A Theoretical Construct for Interpreting Photographs," *Studies in Art Education* 27, no. 2 (January 1, 1986): 52–60; Leonard Henny, "Trend Report: Theory and Practice of Visual Sociology," in *SAGE Visual Methods*, ed. Jason Hughes (London: SAGE, 2012), 20–21.
16. Martha A. Sandweiss, *Print the Legend: Photography and the American West* (New Haven: Yale University Press, 2002); James R. Ryan, *Picturing Empire: Photography and the Visualization of the British Empire* (Chicago: University of Chicago Press, 1997); Elspeth H. Brown, *The Corporate Eye: Photography and the Rationalization of American Commercial Culture, 1884–1929* (Baltimore: Johns Hopkins University Press, 2005).
17. Denis Cosgrove, "Contested Global Visions: One-World, Whole-Earth, and the Apollo Space Photographs" *Annals of the Association of American Geographers* 84, no. 2 (June 1994): 283.
18. For more on photography's role in conveying a sense of experience, see Susan Sontag, *On Photography* (New York: Picador, 2001), 10.
19. Daniel J. Boorstin, *The Image: A Guide to Pseudo-Events in America* (New York: Vintage Books, 1961), 78.
20. Kathy Keltner, "From Myth to Metaphor to Memory: A Rhetorical Analysis of Televised Representations of Project Apollo, 1968–2004," Dissertation (Ohio University, 2007); Anne Collins Goodyear, "The Relationship of Art to Science and Technology in the United States, 1957–1971: Five Case Studies," Dissertation (University of Texas at Austin, 2002).
21. W. J. T. Mitchell, *Picture Theory: Essays on Verbal and Visual Representation* (Chicago: University of Chicago Press, 1994), 11.
22. John Tagg, "Evidence, Truth, and Order: A Means of Surveillance," in *Visual Culture: The Reader*, eds. Jessica Evans and Stuart Hall (London: Sage in association with the Open University, 1999), 244–45.
23. Roger D. Launius, "Public Opinion Polls and Perceptions of US Human Spaceflight," *Space Policy* 19, no. 3 (August 2003): 163–75.
24. David Lubin, *Shooting Kennedy: JFK and the Culture of Images* (Berkeley: University of California Press, 2003), xii.
25. Richard W. Underwood, interviewed by Summer Chick Bergen, October 17, 2000, interview transcript, NASA Johnson Space Center Oral History Project, Houston, TX, 19.
26. Lazier, "Earthrise: or, The Globalization of the World Picture," 627.

27. Maurice Halbwachs, *On Collective Memory*, trans. Lewis Coser (Chicago: University of Chicago Press, 1992).
28. Pierre Nora, "Between Memory and History: les Lieux de Memoire," *Representations* 26 (1989): 12.
29. On the significance of media use of photographs and forming public memory, see Joanne Garde-Hansen, *Media and Memory* (Oxford: Oxford University Press, 2011).
30. For a selection of scholarship since Halbwachs that problematize his theories further, see Michael Billig, "Collective Memory: Ideology and British Royal Family," in *Collective Remembering*, eds. David Middleton and Derek Edwards (London: Sage Publications, 1990); Jan Assman and John Czaplicka, "Collective Memory and Cultural Identity," *New German Critique* 65 (April 1, 1995): 125–33; Alon Confino, "Collective Memory and Cultural History: Problems of Method," *The American Historical Review* 102, no. 5 (December 1, 1997): 1386–1403.
31. Alison Landsberg, *Prosthetic Memory: The Transformation of American Remembrance in the Age of Mass Culture* (New York: Columbia University Press, 2004), 19–20.
32. David Meerman Scott and Richard Jurek, *Marketing the Moon: The Selling of the Apollo Lunar Program* (Cambridge: MIT Press, 2014), 34–53.
33. A memorable use of visuals of Apollo in marketing was a 2011 advertisement from Plantronics, makers of astronaut headsets, which posed the question, "What if our wireless headset wasn't available for those first steps on the Moon?" The visuals follow a "Neil Armstrong" having to use a corded phone to say his famous first words, which are cut off when the cord pulls him back toward the lunar module.

CHAPTER 1: WHY AN AMATEUR NEEDS A BETTER CAMERA THAN A PROFESSIONAL

1. Hasselblad camera advertisement, "Why an amateur needs a better camera than a professional," *Popular Photography* 49, no. 1 (July 1961): inside back cover, 145.
2. While people had certainly seen these places before, the clear view of them from space offered by astronaut photography was unique and new. Underwood interview, NASA, 5.
3. John R. Brinkmann, interviewed by Summer Chick Bergen, March 16, 2001. Interview transcript, NASA Johnson Space Center Oral History Project, Houston, TX, 40–41.
4. Roger D. Launius, "Why Go to the Moon? The Many Faces of Lunar Policy," *Acta Astronautica* 70 (January 2012): 165–75.
5. Matthew H. Hersch, *Inventing the American Astronaut* (New York: Palgrave Macmillan, 2012), 44–45.
6. Brinkmann, interview transcript, 4.
7. Eugene Edmonds, interviewed by Sandra Johnson, November 25, 2003. Interview transcript, NASA Johnson Space Center Oral History Project, Houston, TX, 6.
8. William P. Taub, interviewed by Sandra Johnson, November 8, 2006. Interview transcript, NASA Johnson Space Center Oral History Project, Houston, TX, 9–10.
9. Brinkmann, interview transcript, 6–7.
10. For more on the overall history of the space race, see: Martin J. Collins, *Space Race: The U.S.-U.S.S.R. Competition to Reach the Moon* (San Francisco: Pomegranate, 1999).

11. The Manned Spacecraft Center was renamed the Lyndon B. Johnson Space Center in 1973 for the former president and champion of the spaceflight cause in Texas, "'Johnson Space Center' Name Now Official," *Los Angeles Times*, February 20, 1973, Part I, A14.
12. Brinkmann, interview transcript, 15.
13. John H. Glenn and M. Scott Carpenter, interview with the author, June 26, 2011, transcript, 3.
14. Kraft and Carpenter traded barbs about the reasons for the overshoot of the landing site in their autobiographies, Kraft claiming it was too much photography and a "malfunctioning" astronaut, while Carpenter blamed delayed firing and underperforming retrorockets: Christopher C. Kraft, *Flight: My Life in Mission Control* (New York: Dutton, 2001), 164–70; M. Scott Carpenter and Kris Stoever, *For Spacious Skies: The Uncommon Journey of a Mercury Astronaut* (Orlando: Harcourt, 2002), 286.
15. John H. Boynton and E. M. Fields, "Spacecraft and Launch-Vehicle Performance," in *Results of the Second U.S. Manned Orbital Space Flight, May 24, 1962*, NASA SP-6 (Washington, D.C.: NASA, 1962), 13.
16. Hasselblad, "Hasselblad in Space," https://www.hasselblad.com/history/hasselblad-in-space, accessed 8/28/2019; Walter M. Schirra, Jr. and Richard N. Billings, *Schirra's Space* (Boston: Quinlan Press, 1988), 80; Walter M. Schirra, Jr, interviewed by Roy Neal, December 1, 1998, interview transcript, NASA Johnson Space Center Oral History Project, Houston, TX, 18.
17. John H. Boynton and Lewis R. Fisher, "Spacecraft and Launch-Vehicle Performance," in *Results of the Third United States Manned Orbital Space Flight, October 3, 1962*, NASA SP-12 (Washington, D.C.: NASA, 1962), 9.
18. Schirra, *Schirra's Space*, 80.
19. Lora Bartman, "Photographer Built Camera for Glenn's Flight," *Spaceport News* 37, no. 4 (February 27, 1998): 7.
20. Pan American World Services was under contract to the U.S. Air Force at Patrick AFB to run some operations on the base, and apparently did maintain some kind of photographic lab operations as well. Schirra, interview transcript, 19.
21. At their November 2014 space auction, RR Auctions of Boston, Massachusetts, sold a camera previously owned by Gordon Cooper. My own visual inspection of contemporary photographs of the camera and the camera itself revealed this camera's lens was used on both missions, but the camera body appears to be that of the *Sigma 7* flight. It appears Cooper and Schirra may have used different camera bodies, and Schirra sold what he believed to be his Mercury Hasselblad at auction in 1994.
22. Gordon Cooper, *Leap of Faith: An Astronaut's Journey Into the Unknown* (New York: HarperCollins, 2002), 49.
23. Cooper, *Leap of Faith*, 49.
24. National Aeronautics and Space Administration, *The Triumph of Astronaut L. Gordon Cooper, Jr. and the Faith 7*, Report #NASA-TM-85532 (Houston: NASA, 1963), 19.
25. Up until 2008, this camera was not identified as the Robot camera Cooper used on *Faith 7*. Documentation from the Museum's registrarial files, consisting largely of transfer documents from NASA, does not note it as such. Following additional research into the camera in mission documents, the added feet, stripped exterior, loss of the original viewfinder, and addition of the larger advance dial make this camera, in all probability, one of two still cameras Cooper used on his Mercury mission.

26. Glenn/Carpenter, interview transcript, 3.
27. See examples of this in Frank Van Riper, "Space Cameras: From Ansco to Hassy to Digital." *Washington Post*, October 23, 2003; Robert Pearlman, "Sigma 7 at 50: Retro Space Images Recall 5th U.S. Spaceflight," *Space.com*, October 3, 2012.
28. Glenn/Carpenter, interview transcript, 6.
29. Taub, interview transcript, 9.
30. Taub, interview transcript, 12.
31. The most detailed work, which is more about technology than serious historical examination, is Dwight Steven-Boniecki, *Live TV From the Moon* (Burlington, ON: Apogee Books, 2010). Major works on space technology such as McDougall, *. . . the Heavens and the Earth: A Political History of the Space Age* do not mention cameras other than in a passing fashion. Previously mentioned works by Robert Poole and Benjamin Lazier consider images on a scholarly level, but with an aim to speak to intellectual historians.
32. Interestingly, the book *Dark Moon* aimed to refute the truth of the lunar landings with image analysis in the first chapter, "Photo Call." The authors spend nearly all of the first one hundred pages detailing the supposed lies perpetrated by NASA, the astronauts, and engineers from Hasselblad and other corporations involved in making camera equipment for the space program, both in terms of inadequate technology and purposeful handiwork with processing. The supposed scientific nature of their examinations of the photography is laughable and barely worthy of a mention here as it lacks all markers of serious scholarship. Mary Bennett and David Percy, *Dark Moon: Apollo and the Whistle-Blowers* (Kempton, IL: Adventures Unlimited Press, 2001), 7–76.
33. Hasselblad website, https://www.hasselblad.com/history/hasselblad-in-space, accessed August 28, 2019.
34. This event is noteworthy historically since Sweden remained neutral during the war.
35. "The Two New Hasselblads," *Modern Photography,* December 1954, 72–77 and 124–25.
36. Paul Ceruzzi explains the expansion of options in personal computing that opened up in the 1980s, and in software for them not until 1990. Paul E. Ceruzzi, *A History of Modern Computing* (Cambridge: MIT Press, 2003), 9, 96.
37. John R. Brinkmann and John M. Eggleston, "Photographic Requirements for Manned Space Missions," *SPIE Journal*, no. 3 (Feb.–Mar. 1965): 83.
38. Bartman, "Photographer Built Camera for Glenn's Flight."
39. Albert J. Derr, *Photography Equipment and Techniques: A Survey of NASA Developments*, NASA SP-5099 (Washington, D.C.: NASA, 1972), 5.
40. Eastman Kodak Company, "Final Report, Lunar Photo Study, Contract NAS9-3826," October 1, 1965; C. W. Wyckoff and J. C. McCue, *A Study to Determine the Optimum Design of a Photographic Film for the Lunar Surface Hand-Held Camera*, Prepared for NASA by EG&G, Inc., June 2, 1965.
41. There is some evidence, scant though it may be, that NASA also considered a Leica MDa for Apollo, though no documentation has yet turned up to prove such a claim. See this blogpost from February 2019, http://gmpphoto.blogspot.com/2019/02/leica-cameras-used-by-nasa.html, and this auction notice from late 2014, https://www.paulfrasercollectibles.com/blogs/antiques/nasa-leica-mda-camera-achieves-562-500-at-bonhams-hong-kong.
42. Buy American Act of 1933, 41 U.S.C. § 8302, signed by President Herbert Hoover and kept current through Public Law 112–283.

43. For a contemporary perspective on the benefits of involvement in the space program, see Renato Perez, "Prestige is High in Space Photos," *New York Times*, September 25, 1966, 162.
44. Jocelyn R. Gill, Memorandum to Science Mission Director, November 10, 1964, Johnson Space Center History Collection, Record #22741, Location 075-13, University of Houston, Clear Lake Archives, Clear Lake, TX.
45. Gill Memorandum, November 10, 1964, Record #22741, Location 075-13.
46. Brinkmann and Eggleston, "Photographic Requirements for Manned Space Missions," 83–84.
47. Derr, *Photography Equipment and Techniques*, 5.
48. Helmut A. Kuehnel, *Apollo Experience Report—Photographic Equipment and Operations During Manned Space-Flight Programs* (Washington, D.C.: NASA, 1972), 12.
49. "Gemini Program Mission Report: Gemini V," NASA Document MSC-G-R-65-4, October 1, 1965, 8-1 to 8-48.
50. Mission records reflect fewer cameras onboard Gemini V than Cooper recalls, around three or four with a number of special lenses: Cooper, *Leap of Faith*, 129–30; L. Gordon Cooper, Jr., interviewed by Roy Neal, May 21, 1998, interview transcript, NASA Johnson Space Center Oral History Project, Houston, TX, 24.
51. A general notation in the mission report indicates over two hundred hours spent on "briefings," but gives no specific data on what portion of this covered the three experiments carried out. *Gemini III Mission Report*, MSC-G-R-65-2 (Houston: NASA, 1965), 7–44.
52. The numbering system of Gemini missions changed with this flight. The mission report for Gemini IV lacks the tabulated training information found in all of the other mission reports for this program. General comments regarding training briefings indicate that two experiment briefings and a short experiments review were held at the Kennedy Space Center. *Gemini IV Mission Report*, MSC-G-R-65-3 (Houston: NASA, 1965), 7–14.
53. Gemini VIII was cut short due to a thruster malfunction and no experiments were carried out.
54. No time is listed in the mission report Table 7.1.1-I.—Crew Training Summary. *Gemini IX-A Mission Report*, MSC-G-R-66-6 (Houston: NASA, 1966), 7–11. Completion of five of the seven listed experiments would indicate some preflight training. With the loss of prime crew members Elliot See and Charles Bassett in an aircraft crash two and a half months before the scheduled launch meant the backup crew, Thomas Stafford and Eugene Cernan, stepped up to the prime crew positions. Stafford's previous flight on Gemini VI-A, Cernan's focus on the planned extravehicular activity, and previous status as the backup crew may have contributed to poor accounting for their experiment training.
55. Wilbur A. Ballentine, "DoD/NASA Gemini Experiments Summary," in *Gemini Summary Conference Proceedings,* February 1–2, 1967, SP-138 (Washington, D.C.: NASA, 1967), 309.
56. James David, "Astronaut Photography and the Intelligence Community: Who Saw What?" *Space Policy* 22, no. 3 (August 2006): 185–93.
57. By the time of the final mission report, for Gemini XII, tables for photographic data results show a breakdown between experiment photography and other types of photography, with the vast majority of images being general interest and not related to experiment work. *Gemini XII Mission Report*, MSC-G-R-67-1 (NASA: Houston, 1967), 12–26.
58. Vernon C. Melliff, Memorandum for the Record, February 28, 1964, Johnson Space Center History Collection, Record #27201, Location 075-15, University of Houston, Clear Lake

Archives, Clear Lake, TX. See visual examples of their early thinking in Brinkmann and Eggleston, "Photographic Requirements," 87–89.

59. Donald K. Slayton, Memorandum to Robert Gilruth, May 17, 1966, Johnson Space Center History Collection, Record #27201, Location 075-15, University of Houston, Clear Lake Archives, Clear Lake, TX.
60. Paul Purser, Memorandum through George Low to Robert Gilruth. January 24, 1968, Johnson Space Center History Collection, Record #32389, Location 075-24, University of Houston, Clear Lake Archives, Clear Lake, TX.
61. Donald K. Slayton, Document Review Memorandum to Jim Marsh, October 31, 1967, Johnson Space Center History Collection, Record #32389, Location 075-24, University of Houston, Clear Lake Archives, Clear Lake, TX.
62. Paul Purser, Memorandum for the Record, June 13, 1966, Johnson Space Center History Collection, Record #27389, Location 075-15, University of Houston, Clear Lake Archives, Clear Lake, TX.
63. Warren J. North, Memorandum to Systems Engineering Chief (Peckham), May 15, 1968, Johnson Space Center History Collection, Record #33470, Location 069-63, University of Houston, Clear Lake Archives, Clear Lake, TX.
64. Steven-Boniecki, *Live TV From the Moon*, 18.
65. The conflict between astronauts and television work was most infamously experienced during the Apollo 7 mission when the crew refused to complete a live television broadcast because of their packed schedule. Commander Wally Schirra led what some at NASA termed a mutiny by insisting his crew rest and not perform unscheduled broadcasts. Schirra, *Schirra's Space*, 202.
66. Jeff Bremer, email response to interview questions from the author, August 14, 2012: 2.
67. John Brinkmann, Memorandum to Chief, Systems Engineering Division, Through the Director of Administration, May 13, 1968, Johnson Space Center History Collection, Record #33430, Location 069-62, University of Houston, Clear Lake Archives, Clear Lake, TX.
68. Brinkmann Memo, May 13, 1968, Record #33430, Location 069-62, p. 3. Emphasis in original document.
69. Bremer interview, 4.
70. Deke Slayton, Memorandum to Procurement and Contracts Division, Attn: L.G. Damewood, August 22, 1967, Johnson Space Center History Collection, Record #30818, Location 075-22, University of Houston, Clear Lake Archives, Clear Lake, TX.
71. Wilmot N. Hess, *1967 Summer Study of Lunar Science and Exploration*, NASA SP-157 (Washington, D.C.: NASA, 1967), 398. NASA Historical Research Collection, NASA Headquarters, Folder #012443.
72. Slayton Memo, August 22, 1967, Record #30818, Location 075-22: 7.
73. Bremer interview, 5.
74. Alfred Worden, email response to interview questions, August 10, 2012: 1.
75. Farouk El-Baz, "Training Apollo Astronauts in Lunar Orbital Observations and Photography," in *Analogs for Planetary Exploration*, eds. W. Brent Garry and Jacob E. Bleacher (Boulder, CO: Geological Society of America, 2011), 49–66.
76. James R. Hansen, *First Man: The Life of Neil A. Armstrong* (New York: Simon & Schuster, 2006), 498. In case of an urgent need to leave the lunar surface, plans for astronaut work

on all missions included collecting a contingency soil and rock sample small enough for the astronaut to carry back to the lunar module by hand.

77. Hansen, *First Man*, 498.
78. Hansen, *First Man*, 498.
79. The only note about taking the rising earth photo contained in mission documentation or the recollections of the crew members comes from the first volume of the flight plan, which includes a page for "lunar photo flyby procedures" in the reference materials at the back. It contains a line item for photographing "earth set and earthrise" with the 16mm Maurer moving film camera, not the Hasselblad still photography camera.
80. Richard Allenby, *Analysis of Apollo 8 Photography and Visual Observations*, NASA SP-201 (Washington, D.C.: NASA, 1969).
81. Alan Bean, *Painting Apollo: First Artist on Another World* (Washington, D.C.: Smithsonian Books, 2009), 214–15. Paintings referenced include *The Fabulous Photo We Never Took*, 1999, and *Our Little Secret*, 2003. The scene was also reproduced for television in "That's All There Is," *From the Earth to the Moon,* dir. by Jon Turteltaub (HBO, 1998; HBO Home Video, 2009 DVD).
82. "Apollo 16 Mission Transcript," *Apollo Lunar Surface Journal*, eds. Eric Jones and Ken Glover, Mission Elapsed Time (MET) 120:25:23, http://www.hq.nasa.gov/alsj/a16/a16.html, ALSEP Off-load section, accessed August 28, 2019.
83. Eric Jones and Ken Glover, *Apollo Lunar Surface Journal*, https://www.hq.nasa.gov/alsj. Young stepped onto the lunar surface at approximately MET 119:03:54.
84. Farouk El-Baz, Memorandum for File, "Apollo 10 Photo Debriefing, Case 340," NASA-CR-106888, June 12, 1969: 4.
85. Noel Lamar, Memorandum to Chief, Photographic Technology Laboratory, June 10, 1969, Johnson Space Center History Collection, Record #38181, Location 071-42, University of Houston, Clear Lake Archives, Clear Lake, TX.
86. El-Baz, "Apollo 10 Photo Debriefing, Case 340," 5.
87. Bremer interview, 2.
88. Bill Anders admitted to using ASA 2000, a high-speed film for the Earth-facing side of the Moon, while on the night side of the Moon, an error corrected by special processing by the PTL. William Anders, interviewed by Paul Rollins, October 8, 1997, interview transcript, NASA Johnson Space Center Oral History Project, Houston, TX, 12-14 to 12-15; Bremer recalled one error made by the photo lab, but not the specific mission: Bremer interview, 2. Anecdotal stories circulate about cameras jamming, but nothing substantive appears in mission reports.

CHAPTER 2: PHOTOGRAPHS FOR EVERY AUDIENCE

1. Anders, interview transcript, 12-13 to 12-14.
2. While the Hasselblad Electronic Cameras required manual operation, it was possible to attach an intervalometer and bracket in order to capture timed sequences of still images while passing over the lunar surface. On Apollo missions 8 through 14, this device permitted automatic mapping photography before the addition of a mapping camera in the science section of the service module for the J-type science missions, Apollo 15–17.

3. William Anders, *Apollo 8 Flight Journal*, eds. W. David Woods and Frank O'Brien, mission elapsed time 075:47:30, corrected transcript and commentary, 2009, accessed October 11, 2014.
4. While the immediately accessible and cataloged NASA documentation is incomplete in respect to the entire process of still photography planning and processing, additional research in the papers of upper tier NASA managers may yet reveal more detail on this subject.
5. Helmut Kuehnel, Memorandum for Associate Director, September 19, 1961, Record Group 255, Entry E. 20, Subject File SS-6, Box 2, National Archives and Records Administration, Southwest Center, Fort Worth, TX.
6. Kuehnel Memorandum, September 19, 1961, 2–3.
7. For examples, see final flight plans such as NASA Manned Spacecraft Center, *Flight Plan for MA-7/18*, Langley Station, Hampton, VA, May 1, 1962, Record Group 255, Entry 155, Box 1, National Archives and Records Administration, Southwest Center, Fort Worth, TX. For Apollo objectives and plans, see Richard Orloff, *Apollo by the Numbers: A Statistical Reference* (Washington, D.C.: NASA, 2000).
8. NASA Manned Spacecraft Center, *Flight Plan for MA-9/20*, prepared by Spacecraft Operations Branch, Flight Crew Operations Division, Houston, TX, March 4, 1963, Record Group 255, Entry 155, Box 2, National Archives and Records Administration, Southwest Center, Fort Worth, TX.
9. For an extensive list of Apollo designations, see George Mueller, Memorandum to Distribution, May 2, 1968, Attachment to Record #33720, Location 075-25, University of Houston, Clear Lake Archives, Clear Lake, TX.
10. Experiments Section, Mission Operations Branch, Flight Crew Support Division, *Final Photographic and TV Operations Plan, Apollo 8,* November 18, 1968, Johnson Space Center History Collection, Record #209525, Location 077-66, University of Houston, Clear Lake Archives, Clear Lake, TX.
11. Peter Galison, *Image and Logic: A Material Culture of Microphysics* (Chicago: University of Chicago Press, 1997), 19.
12. These are figures I calculated during extensive evaluations of flight images currently online at websites such as March to the Moon (http://tothemoon.ser.asu.edu). The figures are at best representative of reality but cannot be counted on as absolutely certain.
13. Not included here are either television or movies recorded during the flight. Still frames from the Pilot Observer Cameras were often reproduced as stills in printed publications.
14. *Friendship 7* flight of John Glenn was the first to have a handheld camera for astronaut use, so the flights of Alan Shepard (*Freedom 7*) and Gus Grissom (*Liberty Bell 7*) are not listed here.
15. For Gemini photographic identification, see "Gemini Photography Identification," NASA-TM-110543, NASA Manned Spacecraft Center, 87.
16. Apollo 16 and 17 missions used 35mm Nikon cameras in addition to their Hasselblads. Those images are included in the figures here.
17. Figures available in photo index document produced for each flight.
18. Henry Goett, Forwarded letter from Dr. Jocelyn Gill to Dr. John O'Keefe, Goddard Spaceflight Center, dated December 13, 1962, received by Gilruth on January 21, 1963, Record Group 255, Subject File SS-1.1, File 0627, National Archives and Records Administration, Southwest Center, Fort Worth, TX.

19. Brian Duff, Memorandum to James Webb, November 18, 1963, Office of Public Affairs, 1958–1968, Correspondence, NASA Historical Research Collection, NASA Headquarters, Folder #18171, Box 1.
20. Willis Foster, Memorandum to MSC Director Robert Gilruth, December 3, 1964, Johnson Space Center History Collection, Record #22903, Location 075-13, University of Houston, Clear Lake Archives, Clear Lake, TX.
21. Underwood, interview transcript, 5.
22. George Mueller, Memorandum to the Associate Administrator for OSSA, Homer Newell, March 3, 1966, Johnson Space Center History Collection, Record #26650, Location 066-55, University of Houston, Clear Lake Archives, Clear Lake, TX.
23. Milton Rosen, Memorandum for the Record, November 1, 1967, Johnson Space Center History Collection, Record #31485, Location 068-65, University of Houston, Clear Lake Archives, Clear Lake, TX.
24. Julian West, Memorandum for the Record, December 17, 1968, Johnson Space Center History Collection, Record #36066, Location 075-31, University of Houston, Clear Lake Archives, Clear Lake, TX.
25. General Samuel C. Phillips, Memorandum to Robert Gilruth, March 29, 1968, Johnson Space Center History Collection, Record #32984, Location 069-51, University of Houston, Clear Lake Archives, Clear Lake, TX.
26. Gill Memorandum, November 10, 1964.
27. Deke Slayton, Memorandum to Apollo Spacecraft Program Office Manager, October 26, 1964, Johnson Space Center History Collection, Record #22623, Location 075-13, University of Houston, Clear Lake Archives, Clear Lake, TX.
28. Spacecraft Operations Branch, Flight Crew Support Division, *Preliminary GT-5 Flight Crew Debriefing Transcript Part I*, September 1, 1965, Record Group 255, Entry 155, Box 10, National Archives and Records Administration, Southwest Center, Fort Worth, TX.
29. *Preliminary GT-5 Flight Crew Debriefing Transcript Part I*, 43.
30. George Low, Memorandum to Paul Purser, Special Assistant to the Director, February 21, 1968, Johnson Space Center History Collection, Record #32675, Location 075-25, University of Houston, Clear Lake Archives, Clear Lake, TX.
31. Richard Smith, Memorandum to James McDivitt, March 25, 1971, Johnson Space Center History Collection, Record #43521, Location 073-31, University of Houston, Clear Lake Archives, Clear Lake, TX.
32. Memo refers specifically to the first lunar landing, but engineering priorities changed little in follow-up missions; Charles Klabosh (Operations Branch, FCSD), Memorandum to Chief, Mission Operations Branch (Helmet Kuehnel), February 23, 1967, Record Group 255, Entry 154W, Box 7, Document #CF325-7M-88, National Archives and Records Administration, Southwest Center, Fort Worth, TX.
33. NASA Manned Spaceflight Center, *Flight Plan for MA-8/16*, August 7, 1962, Houston, TX, Record Group 255, Entry 155, Box 1, National Archives and Records Administration, Southwest Center, Fort Worth, TX.
34. NASA Public Affairs Office, *MA-8 Press Conference Transcript*, Houston, TX, October 7, 1962: 10. Record Group 255, Entry 60, Box 2, National Archives and Records Administration, Southwest Center, Fort Worth, TX.

35. Manned Spacecraft Center, *MA-9 Science Debriefing*, June 26, 1963. Record Group 255, Entry 155, Box 2. National Archives and Records Administration, Southwest Center, Fort Worth, TX.
36. Flight Crew Support Division, Spacecraft Operations Branch, *GT-4 Flight Plan*, May 19, 1965, Record Group 255, Entry 155, Box 7, National Archives and Records Administration, Southwest Center, Fort Worth, TX; Gordon Hrabal, *Experiments for GT-4 Mission, NASA Program Gemini Working Paper No. 5023*, May 14, 1965, Record Group 255, Entry 155, Box 7, National Archives and Records Administration, Southwest Center, Fort Worth, TX.
37. Warren North, Memorandum to Manager of the Gemini Program Office, November 4, 1966, Record Group 255, Entry 154W, Box 5, NASA Document #CF323-6M-504, National Archives and Records Administration, Southwest Center, Fort Worth, TX.
38. Only some of the original seven Mercury astronauts and many of those who followed noted in some way their photographic work. The best examples appear in their memoirs. Schirra, *Schirra's Space*. Thomas Stafford and Michael Cassutt, *We Have Capture: Tom Stafford and the Space Race* (Washington, D.C.: Smithsonian Institution Press, 2002), 129. Scott Carpenter and Kris Stoever, *For Spacious Skies: The Uncommon Journey of a Mercury Astronaut* (Orlando: Harcourt, 2002), 211, 263, 295. John Glenn and Nick Taylor, *John Glenn: A Memoir* (New York: Bantam Books, 1999), 250–51. Buzz Aldrin and Ken Abraham, *Magnificent Desolation: The Long Journey Home from the Moon* (New York: Harmony Books, 2009), 39–40.
39. Paul Lowman, James McDivitt, and Edward White, "Terrain Photography on the Gemini IV Mission: Preliminary Report," NASA Technical Note D-3982 (Goddard Space Flight Center, June 1967), 13.
40. For early and late program mention of experiment results for photography experiments such as S005, S006, S011, S029, and objects of opportunity, see "Manned Space Flight Experiments: Gemini V Mission," January 6, 1966, and "Interim Report, Manned Space Flight Experiments, Gemini XII Mission," Report #MSC-TA-R-67-3, August 1967.
41. Rosen Memo, November 11, 1967, 1.
42. Fred Pearce, Memorandum to Distribution List, October 28, 1965, Johnson Space Center History Collection, Record #25820, Location 075-14, University of Houston, Clear Lake Archives, Clear Lake, TX; on science priority of ALSCC, see Wilmot Hess, Memorandum to Deke Slayton and George Low, April 21, 1969, Johnson Space Center History Collection, Record #37545, Location 075-33, University of Houston, Clear Lake Archives, Clear Lake, TX.
43. Wilmot Hess, Memorandum to George Low, June 10, 1968, Johnson Space Center History Collection, Record #33718, Location 070-11, University of Houston, Clear Lake Archives, Clear Lake, TX.
44. Wilmot Hess, Memorandum to George Low, July 17, 1968, Johnson Space Center History Collection, Record #34170, Location 070-21, University of Houston, Clear Lake Archives, Clear Lake, TX; Wilmot Hess, Memorandum to Deke Slayton and George Low, August 21, 1968, Record Group 255, Entry 20, Subject File SS-25, Box 3, National Archives and Records Administration, Southwest Center, Fort Worth, TX; Robert Gilruth, Memorandum to Sam Phillips, July 23, 1968, Johnson Space Center History Collection, Record #34262, Location 075-26, University of Houston, Clear Lake Archives, Clear Lake, TX.

45. Anthony Calio, Memorandum to James McDivitt, March 30, 1970, Johnson Space Center History Collection, Record #41381, Location 072-36, University of Houston, Clear Lake Archives, Clear Lake, TX.
46. Lagrangian points are five distinct locations in the Earth-Sun system (or other two body systems). At those points, a smaller third body could, based on the gravitational equilibrium at these points, maintain its position relative to the Earth and Sun. Gegenschein is the reflection of sunlight off body of dust concentrated at the L2 point, directly opposite Earth from the Sun. Al Worden, email to author, 2.
47. "MA-8 Press Conference Transcript," 10.
48. Robert Gilruth, Memorandum to Fred Singer, February 13, 1963, Record Group 255, Subject File SP-20.1, National Archives and Records Administration, Southwest Center, Fort Worth, TX.
49. John Eggleston, Memorandum to James Bost, Mercury Project Office, Record Group 255, Subject File SP-20.6, National Archives and Records Administration, Southwest Center, Fort Worth, TX.
50. Wilbur Ballentine, "DoD/NASA Gemini Experiments Summary," in *Gemini Summary Conference*, SP-138 (Washington, D.C.: NASA, 1967), 307.
51. Dwayne Day, "Astronauts and Area 51: The Skylab Incident," *The Space Review,* January 9, 2006; Leonard David, "Area 51 Declassified: Documents Reveal Cold War 'Hide-and-Seek'," *Space.com,* November 6, 2013.
52. David, "Astronaut Photography and the Intelligence Community," 185.
53. David, "Astronaut Photography and the Intelligence Community," 185.
54. Mission Evaluation Team, "Apollo 8 Mission Report," NASA Report #NASA-TM-X-66369 (Houston: NASA, 1969).
55. Deke Slayton, Memorandum to George Low, May 13, 1968, Johnson Space Center History Collection, Record #33446, Location 075-25, University of Houston, Clear Lake Archives, Clear Lake, TX.
56. Warren North, Memorandum to James Sasser, September 2, 1969, Record Group 255, Entry 154W5, Box 3, Document #CF342-9M-170, National Archives and Records Administration, Southwest Center, Fort Worth, TX.
57. Warren North, Memorandum to Chief, Mission Planning and Analysis, January 22, 1970, Record Group 255, Entry 154W5, Box 4, Document #CF22-70M-001, National Archives and Records Administration, Southwest Center, Fort Worth, TX.
58. While most frequently called the *Blue Marble* image, NASA only began using this name in recent years. The original caption of the photo refers to it as a "full Earth," but anecdotal evidence supports an entirely different source of the "Blue Marble" name. A children's television program *Big Blue Marble*, which aired from 1974–83 on public television, featured more advanced global concepts than *Sesame Street.* Contemporary news articles about the show's creation describe consultations with NASA astronaut Frank Borman: Tom Shales, "'Big Blue Marble': Picking Up a 'Blue Marble'," *Washington Post*, May 3, 1974, C-1, C-6. The first printed usage by "blue marble" in connection to a NASA story was a quote from journalist David Brinkley for a story about Apollo 10: Clay Gowran, "Apollo Repeats: Service to TV Space-Watchers," *Chicago Tribune*, May 20, 1969, B14.
59. Deke Slayton, Memorandum to Homer Newell, May 16, 1969, Record Group 255, Entry 154W5, Box 2, Document #CF342-9M-85, National Archives and Records Administration, Southwest Center, Fort Worth, TX.

60. A Hycon lunar topographic camera flew on Apollo 13 but was not used until the Apollo 14 mission. James McDivitt, Memorandum to Rocco Petrone, November 22, 1969, Johnson Space Center History Collection, Record #40170, Location 075-41, University of Houston, Clear Lake Archives, Clear Lake, TX.
61. Desire for and documentation of lunar mapping cameras is voluminous, and deserves full scholarly treatment in an entirely separate study. As those photographs stemmed from a remotely operated camera, they fall outside the purview of this work as its focus is hand-held cameras.
62. Worden, email to the author, August 10, 2012.
63. Alfred Morrey (Mission Planning and Analysis), Memorandum to informal distribution list, February 9, 1970. Johnson Space Center History Collection, Record #41078, Location 072-26, University of Houston, Clear Lake Archives, Clear Lake, TX.
64. Farouk El-Baz, Memorandum to Lee Scherer, Lunar Exploration Office, September 2, 1969, Johnson Space Center History Collection, Record #39296, Location 075-35, University of Houston, Clear Lake Archives, Clear Lake, TX; Deke Slayton, Memorandum to George Low, September 15, 1969, Johnson Space Center History Collection, Record #480843, Location 079-15B, University of Houston, Clear Lake Archives, Clear Lake, TX.
65. El-Baz, "Training Apollo Astronauts . . . ," 3.
66. Warren North, Memorandum to James Sasser, April 30, 1970, Record Group 255, Entry 154W5, Box4, Document #CF62-70M-130, National Archives and Records Administration, Southwest Center, Fort Worth, TX.
67. John Brinkmann, Memorandum to Chief, System Engineering Division, May 13, 1968, Johnson Space Center History Collection, Record #33430, Location 069-62, University of Houston, Clear Lake Archives, Clear Lake, TX.
68. Paul Haney, Memorandum to Joseph Shea, April 4, 1966, Johnson Space Center History Collection, Included in Record #27055, Location 075-15, University of Houston, Clear Lake Archives, Clear Lake, TX.
69. Maxime Faget, Memorandum to Chief, Mission Operations Branch, February 27, 1969, Johnson Space Center History Collection, Included in Record #36956, Location 071-16, University of Houston, Clear Lake Archives, Clear Lake, TX.
70. Paul Gast, Memorandum to Acting Public Affairs Officer, JSC, June 17, 1971, Johnson Space Center History Collection, Included in Record #43848, Location 075-46, University of Houston, Clear Lake Archives, Clear Lake, TX.
71. Gast Memorandum, June 17, 1971.
72. Charles Grant Memorandum to Jerome Rosenberg, NASA HQ, July 15, 1963, Record Group 255, Folder SP-20.1, National Archives and Records Administration, Southwest Center, Fort Worth, TX.
73. Jack Stalder, Director of Programs, ITEK, Memorandum to Walter Williams, NASA Associate Director, June 28, 1963, Record Group 255, Folder SS-3, National Archives and Records Administration, Southwest Center, Fort Worth, TX.
74. Pamela Mack, *Viewing the Earth: The Social Construction of the Landsat Satellite System* (Cambridge: MIT Press, 1990), 20.
75. John Brinkmann, Memorandum to Wesley Hjornevik, Director of Administration, June 24, 1969, Johnson Space Center History Collection, Included in Record #38393, Location 071-44, University of Houston, Clear Lake Archives, Clear Lake, TX.

76. CBS, "A Conversation about the U.S. Space Program with Former President Lyndon B. Johnson," 3.
77. Richard Underwood, Memorandum to Wesley Hjornevik, Director of Administration, through John Brinkmann, January 29, 1968, Johnson Space Center History Collection, Included in Record #36525, Location 070-64, University of Houston, Clear Lake Archives, Clear Lake, TX.
78. NASA sent nearly all returned crews, and some spacecraft, on goodwill tours around the country and world. Most noteworthy are those examined by Teasel Muir-Harmony in her dissertation, which examines the global diplomatic implications of tours, exhibits, and other outreach performed by NASA: Teasel Muir-Harmony, *Project Apollo, Cold War Diplomacy and the American Framing of Global Interdependence,* Dissertation, MIT, 2014.
79. Kathy Keltner-Previs' dissertation on the television presentation of Apollo provides details on the earliest goals of the PAO being one of service to journalists because of their considerable ability to get messages to Congress and the public.
80. Donald Arabian, Memorandum to John Brinkmann, December 11, 1968, Johnson Space Center History Collection, Record #36006, Location 075-31, University of Houston, Clear Lake Archives, Clear Lake, TX.
81. Arabian Memorandum, December 11, 1968, 1.
82. David, "Astronaut Photography and the Intelligence Community."
83. David, "Astronaut Photography and the Intelligence Community," 5.
84. William Forrester, Memorandum to the Biomedical Branch, October 12, 1966, Johnson Space Center History Collection, Record #28191, Location 076-13, University of Houston, Clear Lake Archives, Clear Lake, TX.
85. Terry Slezak, interviewed by Rebecca Wright, July 29, 2009, interview transcript, NASA Johnson Space Center Oral History Project, Houston, TX, 23–24.
86. Slezak, interview transcript, 23.
87. John Brinkmann, Memorandum to Wesley Hjornevik, November 27, 1968, Johnson Space Center History Collection, Record #35851, Location 070-52, University of Houston, Clear Lake Archives, Clear Lake, TX.
88. The NASA Technical Reports Server (http://ntrs.nasa.gov), available to researchers in public and private (password protected, secure access) modes, acts as the clearinghouse for such materials.
89. Underwood, interview transcript, 24.
90. Slezak, interview transcript, 20.

CHAPTER 3: IMAGES OF EXPLORATION

1. John Thomson, "Photography and Exploration," *Proceedings of the Royal Geographical Society and Monthly Record of Geography* 13, no. 11 (November 1891): 669–70.
2. The term *moonman* became linked to the photo of Aldrin thanks in part to the design created for MTV's Video Music Awards statue in the early 1980s. From its midnight launch in 1981, MTV aligned itself with NASA visual culture, including the first Space Shuttle launch just months before its own. The original opening graphic sequence for the network's hourly programming (i.e., station identification) featured Apollo imagery with added neon

graphics. MTV and design firm Manhattan Design sought to link their story with Apollo as "uncharted territory." They parlayed this visual association into the awards statuette in 1984, while pop artist Andy Warhol followed up with a similar "Moonwalk" series in 1987: see Rob Tannenbaum and Craig Marks, *I Want My MTV: The Uncensored Story of the Music Video Revolution* (New York: Penguin, 2011).

3. Videos with experts annotating these images, of the Wright Flyer and *Moonman*, are available as part of the Smithsonian's Learning Lab: https://learninglab.si.edu/collections/the-wright-stuff-flying-the-wright-flyer/Mu1DdnXPXWqnXmnj#r/92256 and https://www.youtube.com/watch?v=aLe_21kLAcA.
4. Ron Schick and Julia Van Haaften, *The View from Space: American Astronaut Photography, 1962–1972* (New York: C. N. Potter, 1988), 100.
5. W. J. T. Mitchell, *Landscape and Power*, 2nd ed. (Chicago: University of Chicago Press, 2002), 7.
6. Excellent examples of guides from the period of exploring the American West and Antarctic with cameras are: Sir William de Wiveleslie Abney, *Instruction in Photography* (London: Piper & Carter, 1882); George H. Paltridge, *Photographic Instruction Text: A Systematic Course and Working Guide in All the Processes Which Ordinarily Take Up the Attention of Camera Workers* (New York: Photo-Text Press, 1900).
7. For scholarship on the origins of and the so-called heroic nature of polar exploration in the early twentieth century, see: L. B. Quartermain, *South to the Pole: The Early History of the Ross Sea Sector, Antarctica* (Oxford: Oxford University Press, 1967); T. H. Baughman, *Before the Heroes Came: Antarctica in the 1890s* (Lincoln: University of Nebraska Press, 1999); Edward J. Larson, *An Empire of Ice: Scott, Shackleton, and the Heroic Age of Antarctic Science* (New Haven: Yale University Press, 2011); and Alec Wilkinson, *The Ice Balloon: S.A. Andree and the Heroic Age of Artic Exploration* (New York: Knopf Doubleday, 2013).
8. Charles A. Hill, "The Psychology of Rhetorical Images," in *Defining Visual Rhetorics*, eds. Charles Hill and Marguerite Helmers (Mahwah, NJ: Lawrence Erlbaum, 2004), 33.
9. Hill, "The Psychology of Rhetorical Images," 38.
10. William H. Goetzmann, *Exploration and Empire: The Explorer and the Scientist in the Winning of the American West* (New York: Knopf, 1966), xiii.
11. Eva Respini, *Into the Sunset: Photography's Image of the American West* (New York: The Museum of Modern Art, 2009), 12, 30.
12. Joseph Henry, *The Papers of Joseph Henry, Vol. 11*, Doc. 142 (Washington, D.C.: Smithsonian Institution Press, 2007), 291–92.
13. Sandweiss, *Print the Legend*, 181.
14. Sandweiss, *Print the Legend*, 194.
15. Elizabeth C. Childs, "Time's Profile: John Wesley Powell, Art, and Geology at the Grand Canyon," *American Art* 10, no. 1 (Spring 1996): 7.
16. Walter Benjamin, "The Work of Art in the Age of Mechanical Reproduction," in *Visual Culture: The Reader*, eds. Jessica Evans and Stuart Hall (London: Sage, 1999), 72.
17. Sandweiss, *Print the Legend*, 91.
18. William Mitchell explains that finding truth in photographs in the predigital period of photography linked the ideas of the technology and the scientific method as both required an understanding of evidence. William J. Mitchell, *The Reconfigured Eye: Visual Truth in the Post-Photographic Era* (Cambridge: MIT Press, 1994), 23.
19. Sandweiss, *Print the Legend*, 327.

20. Richard White, *"It's Your Misfortune and None of My Own": A History of the American West* (Norman: University of Oklahoma Press, 1991), 121–22.
21. Donald Worster, *A River Running West: The Life of John Wesley Powell* (Oxford: Oxford University Press, 2001), 241.
22. Toby Jurovics, ed., *Framing the West: The Survey Photographs of Timothy H. O'Sullivan* (Washington, D.C.: Library of Congress, 2010), 19, 25.
23. Peter B. Hales, *William Henry Jackson and the Transformation of the American Landscape* (Philadelphia: Temple University Press, 1988), 78.
24. Hales, *William Henry Jackson and the Transformation of the American Landscape*, 27.
25. Caroline Alexander, *The Endurance: Shackleton's Legendary Antarctic Expedition* (New York: Knopf, 1998), 4.
26. Janis L. Edwards, "Echoes of Camelot: How Images Construct Cultural Memory Through Rhetorical Framing," in *Defining Visual Rhetorics*, eds. Charles Hill and Marguerite Helmers (Mahwah, NJ: Lawrence Erlbaum, 2004), 179.
27. Michael Robinson, *The Coldest Crucible: Arctic Exploration and American Culture* (Chicago: University of Chicago Press, 2006), 163.
28. Max Jones, *The Last Great Quest: Captain Scott's Antarctic Sacrifice* (Oxford: Oxford University Press, 2003), 182.
29. Harry John Philip Arnold, *Photographer of the World: The Biography of Herbert Ponting* (Teaneck, NJ: Fairleigh Dickinson University Press, 1971), 33.
30. Cook claimed he reached the North Pole in April 1908 with only two witnesses and a photograph as proof. Peary followed a year later with his own claim, but with no photographic or navigational evidence but five expedition members to verify the team got within five miles of the Pole.
31. Alasdair McGregor, *Frank Hurley: A Photographer's Life* (New York: Viking, 2004), 19–20.
32. Stephanie L. Barczewski, *Antarctic Destinies: Scott, Shackleton and the Changing Face of Heroism* (London: Hambledon Continuum, 2007), 87.
33. Roland Huntford, *Shackleton* (New York: Carroll & Graf, Inc., 1998), 365.
34. Huntford, *Shackleton*, 372.
35. John MacKenna, *Shackleton: An Irishman in Antarctica* (Madison: University of Wisconsin Press, 2002), 128.
36. Alexander, *The Endurance*, 10.
37. Barczewski, *Antarctic Destinies*, 213–14.
38. For more on the continuing story of the photographs, see *The Endurance* website http://wwwse.kodak.com/US/en/corp/features/endurance/home/index.shtml, accessed August 28, 2019.
39. Limerick, *The Legacy of Conquest*, 88.
40. Childs, "Time's Profile," 15.
41. Limerick, *The Legacy of Conquest*, 88.
42. Roger D. Launius, "Perceptions of Apollo: Myth, Nostalgia, Memory or All of the Above?" *Space Policy* 21, no. 2 (May 2005): 129–39.
43. Elizabeth Kessler, *Picturing the Cosmos: Hubble Space Telescope Images and the Astronomical Sublime* (Minneapolis: University of Minnesota Press, 2012), 5.
44. Angela L. Miller, *The Empire of the Eye* (Ithaca: Cornell University Press, 1993), 10.
45. David Wrobel, *The End of American Exceptionalism: Frontier Anxiety from the Old West to the New Deal* (Tucson: University of Arizona Press, 1993), 145.

46. Limerick, *The Legacy of Conquest*, 75.
47. Barthes, "Rhetoric of the Image," 136.
48. Miles Orvell, *American Photography* (Oxford: Oxford University Press, 2003), 41.
49. Nye, *American Technological Sublime*, xix.
50. Childs, "Time's Profile," 12.
51. Sandweiss, *Print the Legend*, 4.
52. The orientation of the image is subject to the perspective of the astronaut inside the spacecraft, and there are differences between the color and black-and-white versions of this photograph.
53. Ryan, *Picturing Empire,* 19–20.
54. Lindbergh, *Earth Shine*, vii.
55. Neil Maher, "Shooting the Moon," *Environmental History* 9, no. 3 (July 2004): 528.
56. Diana Preston, *A First Rate Tragedy: Robert Falcon Scott and the Race to the South Pole* (Boston: Houghton Mifflin, 1998), 5.
57. Ponting's images, while heralded for their depiction of a tragic experience, were viewed as anticlimactic as they did not show the most famous portions of the journey: Scott and his team's demise. Hurley found the climate for his photography significantly chilled by the war and for a culture of heroes developing from the continental battles—not those of the far reaches of the globe. For more, see Barczewski, *Antarctic Destinies*, 219–24.
58. David, "Astronaut Photography and the Intelligence Community."
59. Michael L. Smith, "Selling the Moon: The U.S. Manned Space Program and the Triumph of Commodity Scientism," in *The Culture of Consumption: Critical Essays in American History, 1880–1980*, eds. Richard Wightman Fox and T. J. Jackson Lears (New York: Pantheon Books, 1983), 177–209.
60. Respini, *Into the Sunset*, 12.
61. Miller, *The Empire of the Eye*, 15.
62. Miller, *The Empire of the Eye*, 18.

CHAPTER 4: THE AFTERLIFE OF ASTRONAUT PHOTOGRAPHY

1. Eugene Cernan, interviewed by Rebecca Wright, December 11, 2007, interview transcript, NASA Johnson Space Center Oral History Project, Houston, TX: 32.
2. For more on the social process by which images are integrated into our personal and collective identities, see Julia Sonnevend, "Iconic Rituals: Towards a Social Theory of Encountering Images," in *Iconic Power: Materiality and Meaning in Social Life*, eds. J. C. Alexander et al. (New York: Palgrave Macmillan, 2012), 220.
3. For an excellent historical look at the complexity and origins of the counterculture movement, see Peter Braunstein and Michael William Doyle, eds., "Historicizing the American Counterculture of the 1960s and '70s," in *Imagine Nation: The American Counterculture of the 1960's and 70's* (New York: Routledge, 2002).
4. Stewart Brand, "Photography Changes Our Relationship to Our Planet," *Click!* Smithsonian Photography Initiative, accessed on March 28, 2014, http://click.si.edu/Story.aspx?story=31.
5. Brand, "Photography Changes . . ."

6. Stewart Brand, ed., *Whole Earth Catalog* (Spring 1969): 7.
7. Fred Turner, *From Counterculture to Cyberculture: Stewart Brand, the Whole Earth Network, and the Rise of Digital Utopianism* (Chicago: University of Chicago Press, 2008), 4.
8. Muir-Harmony, "Project Apollo, Cold War Diplomacy and the American Framing of Global Interdependence."
9. "A Conversation about the U.S. Space Program with Former President Lyndon B. Johnson," broadcast during *Man on the Moon: The Epic Journey of Apollo 11*, CBS Television Network, New York, 21 July 1969: 3. Records of the U.S. Information Agency, RG 306, Entry A1 42, Box 4, National Archives and Records Administration, College Park, MD.
10. Johnson kept a framed copy of *Earthrise* in the Oval Office, which he left for his successor, Richard Nixon. Just over a year after his arrival to the office, the photograph was removed, symbolizing Nixon's departure from supporting Apollo: John Logsdon, *After Apollo? Richard Nixon and the American Space Program* (New York: Palgrave MacMillan, 2015), 282.
11. Robert Hariman and John Louis Lucaites, *No Caption Needed: Iconic Photographs, Public Culture, and Liberal Democracy* (Chicago: University of Chicago Press, 2007), 5.
12. While the authors define the term *iconic* immediately, their only references to NASA images with iconic status are the series of images recalled thanks to video coverage of the Space Shuttle *Challenger* disaster of 1986. Hariman and Lucaites, *No Caption Needed*, 3. Discussion of the term and meaning of *iconic* appear regularly in visual culture journals and scholarly books. Important to note is Keith Moxey, "Visual Studies and the Iconic Turn," *Journal of Visual Culture* 7, no. 2 (2008): 131–46; and Mitchell, *Picture Theory*.
13. Compared to television images of today, Apollo era broadcasts appear fuzzy or grainy, as noted by Kathy Keltner's dissertation, "From Myth to Metaphor to Memory," though at the time were considered very clear. Dick Dubrow, "Well, You've Seen It: Buck Rogers Is Real," *The Evening Star*, Washington, D.C., July 21, 1969, D-5.
14. Poole, *Earthrise*, 28.
15. Mary Bennett and David Percy, *Dark Moon: Apollo and the Whistle-Blowers*, (Kempton, IL: Adventures Unlimited Press, 2001).
16. Polling numbers in support of Moon landings spiked around the Apollo 11 landing, but just barely reached over 50 percent. For useful charts on this, see Launius, "Public Opinion Polls and Perceptions of U.S. Human Spaceflight," 167–68.
17. Harry Kendall, Memorandum to Paul Haney, November 8, 1961, Record Group 255, Folder SS-6, National Archives and Records Administration, Southwest Center, Fort Worth, TX; on the Apollo 11 post-mission tour, see Muir-Harmony, Dissertation, 191–205.
18. "Manned Space Flight Experiments Symposium, Gemini Missions III and IV," iii.
19. "Manned Space Flight Experiments Symposium, Gemini Missions III and IV," 22.
20. "Manned Space Flight Experiments Symposium, Gemini Missions III and IV," 32.
21. Richard Underwood, "Space Photography," in *Gemini Summary Conference, February 1–2, 1967* (Houston: NASA, 1967), 231–90.
22. Jocelyn Gill and Willis Foster, "Science Experiments Summary," in *Gemini Summary Conference, February 1–2, 1967* (Houston: NASA, 1967), 292–93.
23. National Aeronautics and Space Administration, *Earth Photographs from Gemini III, IV, and V*, NASA SP-129 (Washington, D.C.: NASA, 1967).
24. Robert Nugent and Lowell Starr, "Gemini Photography Evaluation," NASA Technical Letter–69, March 1967.

25. Paul Lowman, "Geologic Applications of Orbital Photography," NASA Technical Note D-4155 (Washington, D.C.: NASA, April 1967).
26. C. R. O'Dell, Donald York, and Karl Henize, "Structure of the Barnard Loop Nebula as Determined from Gemini 11 Photographs," *The Astrophysical Journal* 150 (December 1967).
27. Paul Lowman, "Geologic Orbital Photography: Experience from the Gemini Program," *Photogrammetria* 24, no. 3–4 (June 1969): 77–108.
28. Lowman, "Geologic Orbital Photography," 93.
29. H. E. C. van der Meer Mohr, "Geological Interpretation of Hyperaltitude Photographs from Gemini Spacecraft," *Photogrammetria* 24, no. 3–4 (June 1969): 167–74.
30. Frank J. Wobber, "Space Photography: A New Analytical Tool for the Sedimentologist," *Sedimentology* 9 (1967): 265–317.
31. Monem Abdel-Gawad, "Geological Structures of the Red Sea Area Inferred from Satellite Pictures," in *Hot Brines and Recent Heavy Metal Deposits in the Red Sea*, eds. E. T. Degens and D. A. Ross (Berlin: Springer-Verlag, 1969), 25–37.
32. Briefly, some examples of scientific articles using the geological samples are: Apollo Lunar Geology Investigation Team, "Geologic Setting of the Apollo 15 Samples," *Science* 125, no. 4020 (January 28, 1972): 407–15; C. Pieters and T. B. McCord, "Characterization of Lunar Mare Basalt Types. A Remote Sensing Study Using Reflection Spectroscopy of Surface Soils," in *Lunar Science Conference, 7th, Houston, Tex., March 15-19, 1976, Proceedings,* Volume 3 (New York: Pergamon Press, 1976), 2677–90; Patrick Howard Donohue, "Origin and Evolution of High-Titanium Mare Basalts," Diss., University of Notre Dame, 2013.
33. All lunar material remains the property of the U.S. government except for pieces lost/stolen or gifted to those deemed NASA heroes (plaques with their embedded lunar samples are required to be publicly viewable). For the story of one team's quest to locate missing and stolen samples, see Joe Kloc, "The Case of the Missing Moon Rocks," *Atavist* 12 (February 2012).
34. "Apollo 13 Mission: Hearing Before the Committee on Aeronautical and Space Sciences," United States Senate, Ninety-First Congress, Second Session, April 24, 1970 (Washington, D.C.: GPO, 1970).
35. "The Apollo 13 Accident: Hearings Before the Committee on Science and Astronautics," United States House of Representatives, Ninety-First Congress, Second Session, No. 19, June 16, 1970 (Washington, D.C.: GPO, 1970).
36. "Twentieth Semi-Annual Report to Congress, July 1–December 31, 1968," House Document 91–153, October 3, 1969 (Washington, D.C.: NASA, 1969). It should be noted that the vast majority of the report detailed activities other than human spaceflight, reflecting NASA's actual work and not just what garnered the most public attention.
37. "1972 NASA Authorization: Hearings Before the Committee on Science and Astronautics," First Session on H.R. 3981 (superseded by H.R. 7109), March 2, 1971 (Washington, D.C.: GPO, 1971).
38. House Committee on Science and Astronautics, "Apollo 15 Mission Report," HRG-1971-SAH-0005 (September 9, 1971); "Apollo 16 Mission Report," HRG-1972-SAH-0003 (May 16, 1972); "1974 NASA Authorization," HRG-1973-SAH-0023 (March 14, 1973), 501–33.
39. Rob Kroes, *Photographic Memories: Private Pictures, Public Images, and American History* (Hanover, NH: Dartmouth College Press, 2007), 13.

40. Marita Sturken, *Tangled Memories: The Vietnam War, the AIDS Epidemic, and the Politics of Remembering* (Berkeley: University of California Press, 1997).
41. Complete mission coverage on major networks such as CBS occurred for nearly every U.S. human mission from Alan Shepard's Mercury flight through Apollo 12.
42. George Low, Memorandum to Richard Johnston, Special Assistant to Robert Gilruth, March 18, 1969, Johnson Space Center History Collection, Record #37166, Location 071-23, University of Houston, Clear Lake Archives, Clear Lake, TX.
43. Underwood, interview transcript, 19.
44. Edgar Cortwright, Memorandum to Robert Gilruth, March 25, 1969, Johnson Space Center History Collection, Record #37237, Location 071-24, University of Houston, Clear Lake Archives, Clear Lake, TX.
45. Melvin Day, Memorandum to Robert Gilruth, May 26, 1969, Johnson Space Center History Collection, Record #37973, Location 071-36, University of Houston, Clear Lake Archives, Clear Lake, TX.
46. George Low, Memorandum to Robert Gilruth, April 3, 1969, Record #37341, Location 071-25; Richard Johnston, Memorandum to George Low, April 9, 1969, Record #37392, Location 071-25; Richard Johnston, Memorandum to Distribution, April 21, 1969, Record #37531, Location 071-31; all from Johnson Space Center History Collection, University of Houston, Clear Lake Archives, Clear Lake, TX.
47. Warren Burkett, *News Reporting: Science, Medicine, and High Technology* (Ames: Iowa State University Press, 1986), 30. On how this increased reporting reflected back on science education, see Dorothy Nelkin, *Selling Science: How the Press Covers Science and Technology* (New York: W. H. Freeman and Company, 1987), 21.
48. For examples of this practice at different newspapers, see Edwin Diamond, *Behind the Times: Inside the New York Times* (Chicago: University of Chicago Press, 1995); David S. Broder, *Behind the Front Page: A Candid Look at How the News Is Made* (New York: Simon and Schuster, 1987).
49. Howard McCurdy, *Space and the American Imagination* (Washington, D.C.: Smithsonian Institution Press, 1997), 51.
50. Burkett, *News Reporting*, 32.
51. On the intensely competitive nature of print journalists and their insider perspective on the beats they cover, see Broder, *Behind the Front Page*.
52. On the reliance of public and individual memories on the media, though specifically the often-abused ability to use television to reconstruct events, see Andrew Hoskins, "Television and the Collapse of Memory," in *Time & Society* 13, no. 1 (2004): 109–27. For an excellent discussion of the individual nature of the inscriptive format selected for memories, see José van Dijck, "Mediated Memories: Personal Cultural Memory as Object of Cultural Analysis," *Continuum: Journal of Media & Cultural Studies* 18, no. 2 (2004): 261–77.
53. As noted by Sharon Dunwoody in her commentary on essays in Martin W. Bauer and Massimiano Bucchi, eds., *Journalism, Science and Society: Science Communication Between News and Public Relations* 7 (New York: Routledge, 2007), newspaper readership generally at the time of the Apollo flights would have averaged around one per household in the United States (from data at Editor and Publisher Yearbook Online at http://www.editorandpublisher.com).

54. Political scientist Amber Boydstun gives a statistical analysis of news coverage based on the press having an organizational process, being very competitive, and a proclivity toward what she calls the disproportionate information-processing systems. The final framework would explain the high visibility for human spaceflight in the 1960s but the relatively low public support for it: Amber E. Boydstun, *Making the News: Politics, the Media, and Agenda Setting* (Chicago: University of Chicago Press, 2013), 6.
55. The issue of positioning of *Earthrise* still causes debate among those who frequently publish the image. Based on the recent video created for the forty-fifth anniversary of Apollo 8, the photograph was taken from the perspective typical of horizon images familiar from the perspective of one standing on the surface with the Earth rising from beyond a horizontal line. However, comments made by Bill Anders about his position in relation to the Moon and the spacecraft indicate that what he saw looking out the window was Moon to his left and the Earth appearing from behind on the right, a 90-degree clockwise positioning from the typical view. No matter the orientation selected, every viewing angle was possible as astronauts floated in all directions in zero gravity.
56. "First Look at the Apollo 8 Pictures," *Nature* 221 (January 18, 1969): 215–17.
57. Alan Boyle, "40 Years Later, Apollo 17's Blue Marble Leaves a Mark on Our Memory," *Cosmic Blog*, nbcnews.com (December 7, 2012), accessed November 4, 2014; Mitch Tobin, "The Black and Blue Marble: 10 Whole Earth Views from Space," EcoWest.com (September 12, 2013), accessed November 4, 2014.
58. *Time*, December 25, 1972: 30, 33; January 1, 1973: 49–51.
59. *Newsweek*. December 26, 1972: 45–46; January 2, 1973: 38.
60. Neglecting the human spaceflight story from works on liberal culture and politics of the 1960s is common, though Kennedy and Johnson were major advocates of exploration. See Allen J. Matusow, *The Unraveling of America: A History of Liberalism in the 1960s* (New York: Harper & Row, 1984). Other mentions usually include a notation of program milestones in timelines or brief inclusion as "highlights" of the period. See John C. McWilliams, *The 1960s Cultural Revolution* (Westport, CT: Greenwood Press, 2000); John Robert Greene, *America in the Sixties* (Syracuse: Syracuse University Press, 2010). Informal but repeated references to the space program abound. See M. J. Heale, *The Sixties in America: History, Politics and Protest* (Chicago: Fitzroy Dearborn Publishers, 2001).
61. For example, see of those magazines associated with the three most significant missions, Apollo 8 (issues for December 1968 through January 1969), Apollo 11 (June and July 1969), and Apollo 17 (December 1972 to January 1973). Other examples are a *Look* magazine special edition in 1969, priced at $1.25, featuring *New York Times* staff writer text, a reproduction of a Norman Rockwell painting, and color photographs from Apollo 8. A similar issue, for 25 cents more, was published for Apollo 11 later the same year.
62. The purpose of using such film was the higher contrast available with black-and-white film and the application of the reseau plate for photogrammetry. In the case of Apollo 12, when their color television camera broke and they only had black-and-white film magazines with them on their trek to Surveyor III, all images of that historic moment are black-and-white.
63. For examples, see Mission Operations Branch, Flight Crew Support Division, "Apollo 11 Technical Crew Debriefing, Vol. 1" (Houston: NASA, July 31, 1969): 10–9 to 10–10; "Apollo 12 Technical Crew Debriefing," (Houston: NASA, December 1, 1969): 9–14.

64. John Lindsay, *Lunar Stratigraphy and Sedimentology* (Amsterdam: Elsevier Scientific Publishing Group, 1976), 227–83.
65. While a few scientific/robotic missions returned attention to the Moon, including the currently successful Lunar Reconnaissance Orbiter, the vast majority of space research money since has been spent on spacecraft to other planetary bodies and telescopes for observing objects well beyond it.
66. Norman Mailer, *Of a Fire on the Moon* (Boston: Little, Brown, 1970). Contemporary and modern critiques of the value of Project Apollo can be found in many newspaper archives, but historical criticisms are well summarized in Roger Launius, "Interpreting the Moon Landings: Project Apollo and the Historians," *History and Technology* 22, no. 3 (2006): 225–55.
67. Goodyear, "The Relationship of Art to Science and Technology in the United States, 1957–1971: Five Case Studies"; Goodyear, "NASA and the Political Economy of Art, 1962–1974."
68. Frank J. Malina, "On the Visual Fine Arts in the Space Age," *Leonardo* 3, no. 3 (July 1970): 323–25.
69. The specific Catholic iconography of the mosaic likely played a role in its removal when the center transitioned to have a broad culturally diverse goal. I only speculate here about the mural, but with indications of others at places such as Ben & Jerry's in Vermont, an additional survey of such displays would be appropriate in the future.
70. Poole, *Earthrise*; Neil Maher, *Apollo in the Age of Aquarius* (Cambridge: Harvard University Press, 2017).
71. Maher, 92–136.
72. Cosgrove, 271.
73. Kim McQuaid, "Selling the Space Age: NASA and Earth's Environment, 1958-1990," *Environment & History* 12, no. 2 (May 2006): 127–63.
74. Anders, interview transcript, 15–16.

EPILOGUE: CONTINUING RESONANCE

1. Other than the expansive views of Earth offered by this configuration, its location on the ISS positions it for a visual line of sight for operating the remote manipulator system (Canadarm2) during the approach and docking of resupply spacecraft.
2. For an assessment of technological changes that allowed astronauts to share their experiences with those on Earth, see Valerie Neal, *Spaceflight in the Shuttle Era and Beyond: Redefining Humanity's Purpose in Space* (New Haven: Yale University Press, 2017), 86.
3. Valerie Neal, "Bringing Spaceflight Down to Earth: Astronauts and The IMAX Experience®," in *Spacefarers: Images of Astronauts and Cosmonauts in the Heroic Era of Spaceflight*, ed. Michael J. Neufeld (Washington, D.C.: Smithsonian Institution Scholarly Press, 2013), 149–74; Toni Myers, comments to the author, April 4, 2012.
4. Roberta Bondar, a one-time astronaut, uses her website, robertabondar.com, to share her work.
5. Richard Garriott was the first person to use software developed by the Windows on Earth project while in orbit to determine when to photograph the exact locations his father also photographed. For more, see windowsonearth.org.

6. Michael Soluri, "Examining the Iconic and Rediscovering the Photography of Space Exploration in Context to the History of Photography," in *Remembering the Space Age: Proceedings of the 50th Anniversary Conference*, ed. Steven J. Dick (Washington, D.C.: NASA, 2008), 271–339.
7. The short film "The ISS Image Frontier—Making the Invisible Visible" by science filmmaker Christoph Malin compiled video from the ISS, Pettit's Luminance 2012 presentation, and a running time lapse of Pettit's images to make the film. The final product received a great deal of press for its production quality and artistic composition: http://vimeo.com/61083440, accessed October 14, 2014.
8. For Pettit's eloquent description of his interest in and hopes for his imagery, see his speech from the Luminance 2012 conference, http://vimeo.com/51632896, accessed October 14, 2014.
9. Exclusively for images of Earth, see http://eol.jsc.nasa.gov/; http://visibleearth.nasa.gov. For general collections of images, NASA now only updates their main Flickr page, https://www.flickr.com/photos/nasa2explore, though their old pages run by each center are often still available, examples being from: KSC, http://spaceflight.nasa.gov/gallery/index.html; JSC, http://images.jsc.nasa.gov; Dryden, http://www.dfrc.nasa.gov/Gallery/Photo; Marshall, http://mix.msfc.nasa.gov, all accessed on October 16, 2014.
10. Neil deGrasse Tyson, "Launch Keynote: 28th National Space Symposium," Space Foundation, https://www.youtube.com/watch?v=VLzKjxglNyE.
11. Penny4NASA, http://www.penny4nasa.org, seeks to increase NASA's budget to 1 percent of the federal budget in part based on Tyson's belief that American culture thrives most when we have strong support for exploration.

Archival and Bibliographic Sources

ARCHIVAL SOURCES

The age of digital archives made research for this book far more efficient than that of my space history predecessors. While in-person visits were required for the NASA archives at UHCL and NARA Fort Worth, they were guided by search tools on each location's website. Websites listed here are nontraditional archives in terms of their collecting documentation relevant to specific projects, but their interest in saving such documents places them solidly within the category of digital archives. This book depended on both textual and visual source material, and the online resources for images were invaluable as the photos were often supplemented by additional reports elsewhere on those websites. NASA's own image search engines have changed at least three times throughout the writing of this text, so only the websites available upon completion of this manuscript are cited here.

Apollo Flight Journal, David Woods, NASA Headquarters History Office, Washington, D.C. (https://history.nasa.gov/afj/index.html).

Apollo Image Atlas, Regional Planetary Image Facility, Lunar and Planetary Institute, Houston, TX (https://www.lpi.usra.edu/resources/apollo).

Apollo Lunar Surface Journal, Eric M. Jones and Ken Glover, NASA Headquarters History Office, Washington, D.C. (https://www.hq.nasa.gov/alsj).

Gateway to Astronaut Photography of Earth, International Space Station and Earth Science and Remote Sensing Unit, NASA Johnson Space Center, Houston, TX (https://eol.jsc.nasa.gov).

Johnson Space Center History Collection, University of Houston, Clear Lake Archives, Clear Lake, TX (https://historycollection.jsc.nasa.gov/JSCHistoryPortal/history/history_collection.htm).

Johnson Space Center Oral History Project, Johnson Space Center History Office, Houston, TX (https://historycollection.jsc.nasa.gov/JSCHistoryPortal/history/oral_histories/oral_histories.htm).

March to the Moon, NASA Johnson Space Center with Arizona State University, Houston, TX, and Tempe, AZ (http://tothemoon.ser.asu.edu).

NASA Historical Research Collection, NASA Headquarters, Washington, D.C.

NASA Image and Video Library, NASA Marshall Spaceflight Center, Huntsville, AL (https://images.nasa.gov).

NASA Technical Reports Server (NTRS), STI Program, NASA Langley Research Center, Hampton, VA (https://ntrs.nasa.gov).

National Archives and Records Administration, Record Group 255, Fort Worth Federal Records Center, Fort Worth, TX.

BIBLIOGRAPHY

Abdel-Gawad, Monem. "Geological Structures of the Red Sea Area Inferred from Satellite Pictures." In *Hot Brines and Recent Heavy Metal Deposits in the Red Sea*, edited by E. T. Degens and D. A. Ross, 25–37. Berlin: Springer-Verlag, 1969.

Abney, Sir William de Wiveleslie. *Instruction in Photography*. London: Piper & Carter, 1882.

Aldrin, Buzz, and Ken Abraham. *Magnificent Desolation: The Long Journey Home from the Moon*. New York: Harmony Books, 2009.

Alexander, Caroline. *The Endurance: Shackleton's Legendary Antarctic Expedition*. New York: Knopf, 1998.

Allenby, Richard. *Analysis of Apollo 8 Photography and Visual Observations*, NASA SP-201. Washington, D.C.: NASA, 1969.

Apollo Lunar Geology Investigation Team. "Geologic Setting of the Apollo 15 Samples." *Science* 125, no. 4020 (January 28, 1972): 407–15.

Arnold, Harry John Philip. *Photographer of the World: The Biography of Herbert Ponting*. Teaneck, NJ: Fairleigh Dickinson University Press, 1971.

Assmann, Jan, and John Czaplicka. "Collective Memory and Cultural Identity." *New German Critique*, no. 65 (April 1, 1995): 125–33.

Ballentine, Wilbur A. "DoD/NASA Gemini Experiments Summary." *Gemini Summary Conference Proceedings*, SP-138, February 1–2, 1967, Washington, D.C.

Barczewski, Stephanie L. *Antarctic Destinies: Scott, Shackleton and the Changing Face of Heroism*. London: Hambledon Continuum, 2007.

Barrett, Terry. "A Theoretical Construct for Interpreting Photographs." *Studies in Art Education* 27, no. 2 (January 1, 1986): 52–60.

Barthes, Roland. "Rhetoric of the Image." In *The Visual Culture Reader*, edited by Nicholas Mirzoeff, 135–38. London: Routledge, 2002.

Bartman, Lora. "Photographer Built Camera for Glenn's Flight." *Spaceport News*, 37, no. 4 (February 27, 1998): 7.

Batchen, Geoffrey. *Burning with Desire: The Conception of Photography*. Cambridge, MA: MIT Press, 1997.

Bauer, Martin W., and Massimiano Bucchi, eds. *Journalism, Science and Society: Science Communication between News and Public Relations*. New York: Routledge, 2007.

Baughman, T. H. *Before the Heroes Came: Antarctica in the 1890s*. Lincoln: University of Nebraska Press, 1999.

———. *Pilgrims on the Ice: Robert Falcon Scott's First Antarctic Expedition*. Lincoln: University of Nebraska Press, 2008.

Bean, Alan. *Painting Apollo: First Artist on Another World*. Washington, D.C.: Smithsonian Books, 2009.

Benjamin, Walter. *The Work of Art in the Age of Mechanical Reproduction*. New York: Penguin Books Limited, 2008.

Bennett, Mary, and David Percy. *Dark Moon: Apollo and the Whistle-Blowers*. Kempton, IL: Adventures Unlimited Press, 2001.

Billig, Michael. "Collective Memory: Ideology and British Royal Family." In *Collective Remembering*, edited by David Middleton and Derek Edwards, 60–80. London: Sage Publications, 1990.

Bilstein, Roger. *Stages to Saturn: A Technological History of the Apollo/Saturn Launch Vehicles*. Washington, D.C.: NASA, 1996.

Bodnar, John E. *Remaking America: Public Memory, Commemoration, and Patriotism in the Twentieth Century*. Princeton, NJ: Princeton University Press, 1992.

Boorstin, Daniel J. *The Image: A Guide to Pseudo-Events in America*. 1st ed. New York: Vintage Books, 1992.

Bourdieu, Pierre. *Photography: A Middle-Brow Art*. Translated by Shaun Whiteside. Stanford: Stanford University Press, 1990.

———. "The Social Definition of Photography." In *Visual Culture: The Reader*, edited by Jessica Evans and Stuart Hall, 162–80. London: Sage/Open University, 1999.

Boydstun, Amber E. *Making the News: Politics, the Media, and Agenda Setting*. Chicago: University of Chicago Press, 2013.

Boyer, Paul S. *By the Bomb's Early Light: American Thought and Culture at the Dawn of the Atomic Age*. Chapel Hill: University of North Carolina Press, 1994.

Boynton, John H., and E. M. Fields. "Spacecraft and Launch-Vehicle Performance." In *Results of the Second U.S. Manned Orbital Space Flight, May 24, 1962*, NASA SP-6. Washington, D.C.: NASA, 1962.

Boynton, John H., and Lewis R. Fisher. "Spacecraft and Launch-Vehicle Performance." In *Results of the Third United States Manned Orbital Space Flight, October 3, 1962*, NASA SP-12. Washington, D.C.: NASA, 1962.

Brand, Stewart. "Photography Changes Our Relationship to Our Planet." *Click!* Smithsonian Photography Initiative (Spring 1969). Accessed March 28, 2014. http://click.si.edu/Story.aspx?story=31.

———, ed. *Whole Earth Catalog*.

Braunstein, Peter, and Michael William Doyle, eds. "Historicizing the American Counterculture of the 1960s and '70s." In *Imagine Nation: The American Counterculture of the 1960's and 70's*. New York: Routledge, 2002.

Brick, Howard. *Age of Contradiction: American Thought and Culture in the 1960s*. New York: Twayne Publishers, 1998.

Brinkmann, John R., and John M. Eggleston. "Photographic Requirements for Manned Space Missions." *SPIE Journal*, 3 (February–March 1965).

Broder, David S. *Behind the Front Page: A Candid Look at How the News Is Made*. New York: Simon and Schuster, 1987.

Brooks, Courtney, James Grimwood, and Loyd Swenson. *Chariots for Apollo: A History of Manned Lunar Spacecraft*. Washington, D.C.: NASA, 1979.

Brooks, Peter, G. C. Frewer, and G. K. C. Pardoe. *Project Apollo: The Way to the Moon*. New York: American Elsevier Publishing, 1969.

Brown, Elspeth H. *The Corporate Eye: Photography and the Rationalization of American Commercial Culture, 1884–1929*. Baltimore: Johns Hopkins University Press, 2005.

Brown, Joshua. *Beyond the Lines: Pictorial Reporting, Everyday Life, and the Crisis of Gilded Age America*. Berkeley: University of California Press, 2002.

Burke, Peter. *Eyewitnessing: The Uses of Images as Historical Evidence*. Ithaca, NY: Cornell University Press, 2001.

Burkett, Warren. *News Reporting: Science, Medicine, and High Technology*. 1st ed. Ames: Iowa State University Press, 1986.

Byrnes, Mark E. *Politics and Space: Image Making by NASA*. Westport: Praeger, 1994.

Carpenter, M. Scott, and Kris Stoever. *For Spacious Skies: The Uncommon Journey of a Mercury Astronaut*. Orlando: Harcourt, 2002.

Carter, Edward C. *Surveying the Record: North American Scientific Exploration to 1930*. Philadelphia: American Philosophical Society, 1999.

Cassidy, James G. *Ferdinand V. Hayden: Entrepreneur of Science*. Lincoln: University of Nebraska Press, 2000.

Ceruzzi, Paul E. *A History of Modern Computing*. Cambridge: MIT Press, 2003.

Chaikin, Andrew. *A Man on the Moon: The Voyages of the Apollo Astronauts*. New York: Viking, 1994.

Childs, Elizabeth C. "Time's Profile: John Wesley Powell, Art, and Geology at the Grand Canyon." *American Art* 10, no. 1 (Spring 1996): 7–35.

Collins, Martin J. *Space Race: The U.S.–U.S.S.R. Competition to Reach the Moon*. San Francisco: Pomegranate, 1999.

Confino, Alon. "Collective Memory and Cultural History: Problems of Method." *The American Historical Review* 102, no. 5 (December 1, 1997): 1386–1403.

Cooper, Gordon. *Leap of Faith: An Astronaut's Journey into the Unknown*. New York: HarperCollins, 2002.

Cosgrove, Denis. *Apollo's Eye: A Cartographic Genealogy of the Earth in the Western Imagination*. Baltimore: Johns Hopkins University Press, 2001.

———. "Contested Global Visions: One-World, Whole-Earth, and the Apollo Space Photographs." *Annals of the Association of American Geographers* 84, no. 2 (June 1994): 270–94.

Cowan, Ruth Schwartz. "The Consumption Junction: A Proposal for Research Strategies in the Sociology of Technology." In *The Social Construction of Technological Systems: New Directions in the Sociology and History of Technology*, edited by Wiebe E. Bijker, et al., 253–72. Cambridge: MIT Press, 2012.

Crane, Susan A. "Memory, Distortion, and History in the Museum." *History and Theory* 36, no. 4 (December 1997): 44–63.

Crary, Jonathan. *Techniques of the Observer: On Vision and Modernity in the Nineteenth Century*. October Books. Cambridge: MIT Press, 1992.

David, James. "Astronaut Photography and the Intelligence Community: Who Saw What?" *Space Policy*, 22, no. 3 (August 2006): 185–93.

David, Leonard. "Area 51 Declassified: Documents Reveal Cold War 'Hide-and-Seek'." *Space .com*. November 6, 2013.

David, Robert G. *The Arctic in the British Imagination, 1818–1914*. Manchester: Manchester University Press, 2000.

Day, Dwayne. "Astronauts and Area 51: The Skylab Incident." *The Space Review*. January 9, 2006.

Debord, Guy. "The Society of the Spectacle." In *The Visual Culture Reader*, edited by Nicholas Mirzoeff, 142–44. London: Routledge, 2002.

Derr, Albert J. *Photography Equipment and Techniques: A Survey of NASA Developments*. NASA SP-5099. Washington, D.C.: NASA, 1972.

DeVorkin, David H. *Hubble Imaging Space and Time*. Washington, D.C.: Smithsonian Institution Press, 2008.

Diamond, Edwin. *Behind the Times: Inside the New New York Times*. Chicago: University of Chicago Press, 1995.

Dikovitskaya, Margarita. *Visual Culture: The Study of the Visual After the Cultural Turn*. Cambridge, MA: MIT Press, 2005.

Donohue, Patrick Howard. "Origin and Evolution of High-Titanium Mare Basalts." Dissertation, University of Notre Dame, 2013.

Doss, Erika Lee. *Looking at Life Magazine*. Washington, D.C.: Smithsonian Institution Press, 2001.

Dubrow, Dick. "Well, You've Seen It: Buck Rogers Is Real." *The Evening Star*, July 21, 1969.

Eastman Kodak Company. "Final Report, Lunar Photo Study, Contract NAS9-3826." October 1, 1965.

Edwards, Elizabeth. *The Camera as Historian: Amateur Photographers and Historical Imagination, 1885–1918*. Durham, NC: Duke University Press, 2012.

Edwards, Janis L. "Echoes of Camelot: How Images Construct Cultural Memory Through Rhetorical Framing." In *Defining Visual Rhetorics*, edited by Charles Hill and Marguerite Helmers, 179–94. Mahwah, NJ: Lawrence Erlbaum, 2004.

El-Baz, Farouk. "Training Apollo Astronauts in Lunar Orbital Observations and Photography." In *Analogs for Planetary Exploration*, edited by W. Brent Garry and Jacob E. Bleacher, 49–66. Boulder, CO: Geological Society of America, 2011.

Evans, Jessica, and Stuart Hall, eds. *Visual Culture: The Reader*. London: Sage/Open University, 1999.

Foucault, Michel. "Panopticism." In *Visual Culture: The Reader*, edited by Jessica Evans and Stuart Hall, 61–71. London: Sage/Open University, 1999.

French, Francis, and Colin Burgess. *In the Shadow of the Moon: A Challenging Journey to Tranquility, 1965–1969*. Lincoln: University of Nebraska Press, 2007.

Freud, Sigmund. *The Standard Edition of the Complete Psychological Works of Sigmund Freud*. London: Hogarth Press, 1953.

Freund, Gisèle. *Photography & Society*. Boston: D. R. Godine, 1980.

Freytag, Heinrich. *The Hasselblad Way; the Hasselblad Photographer's Companion*. 5th ed. London, New York: Focal Press, 1973.

Friedman, Sharon M., Sharon Dunwoody, and Carol L. Rogers, eds. *Scientists and Journalists: Reporting Science as News*. New York; London: Free Press; Collier Macmillan, 1986.

Frisch, Michael. "American Memory and the Structures of Collective Memory: A Modest Exercise in Empirical Iconography." *Journal of American History* 75, no. 4 (March 1989): 1130–155.

Galison, Peter. *Image and Logic: The Material Culture of Microphysics*. Chicago: University of Chicago Press, 1997.
Garde-Hansen, Joanne. *Media and Memory*. Edinburgh: Edinburgh University, 2011.
Gill, Jocelyn, and Willis Foster, "Science Experiments Summary." In *Gemini Summary Conference, February 1–2, 1967*. Houston: NASA, 1967.
Glenn, John, and Nick Taylor. *John Glenn: A Memoir*. New York: Bantam Books, 1999.
Goetzmann, William H. *Exploration and Empire; the Explorer and the Scientist in the Winning of the American West*. 1st ed. New York: Knopf, 1966.
Goldberg, Vicki. *The Power of Photography: How Photographs Changed Our Lives*. St. Louis: San Val, 1993.
Goodyear, Anne Collins. "NASA and the Political Economy of Art, 1962–1974." In *The Political Economy of Art: Making the Nation of Culture*, edited by Julie F. Codell, 191–206. Associated University Press, 2008.
———. "The Relationship of Art to Science and Technology in the United States, 1957–1971: Five Case Studies." PhD dissertation, University of Texas at Austin, 2002.
Gowran, Clay. "Apollo Repeats: Service to TV Space-Watchers." *Chicago Tribune*, May 20, 1969.
Greene, John Robert. *America in the Sixties*. Syracuse: Syracuse University Press, 2010.
Hacker, Barton C. *On the Shoulders of Titans: A History of Project Gemini*. Washington: NASA: U.S. Government Printing Office, 1977.
Halbwachs, Maurice. *On Collective Memory*. Chicago: University of Chicago Press, 1992.
Hales, Peter B. "Imagining the Atomic Age." In *Looking at Life Magazine*, edited by Erika Lee Doss, 103–19. Washington, D.C.: Smithsonian Institution Press, 2001.
———. *William Henry Jackson and the Transformation of the American Landscape*. Philadelphia: Temple University Press, 1988.
Handberg, Roger. *Reinventing NASA: Human Space Flight, Bureaucracy, and Politics*. Santa Barbara: Greenwood Publishing Group, 2003.
Hansen, James R. *First Man: The Life of Neil A. Armstrong*. New York: Simon & Schuster, 2006.
Hariman, Robert, and John Louis Lucaites. *No Caption Needed: Iconic Photographs, Public Culture, and Liberal Democracy*. Chicago: University of Chicago Press, 2007.
Hasselblad Camera Advertisement, "Why an Amateur Needs a Better Camera Than a Professional." *Popular Photography* 49, no. 1 (July 1961): inside back cover, 145.
Hawkins, Stephanie L. *American Iconographic: National Geographic, Global Culture, and the Visual Imagination*. Charlottesville: University of Virginia Press, 2010.
Heale, M. J. *The Sixties in America: History, Politics and Protest*. Chicago: Fitzroy Dearborn Publishers, 2001.
Henny, Leonard. "Trend Report: Theory and Practice of Visual Sociology." In *SAGE Visual Methods*, edited by Jason Hughes, 1–64. London: SAGE, 2012.
Henry, Joseph. *The Papers of Joseph Henry*. Washington, D.C.: Smithsonian Institution Press, 1972.
Hersch, Matthew H. *Inventing the American Astronaut*. New York: Palgrave Macmillan, 2012.
Hill, Charles A. "The Psychology of Rhetorical Images." In *Defining Visual Rhetorics*, edited by Charles Hill and Marguerite Helmers, 25–40. Mahwah, NJ: Lawrence Erlbaum, 2004.
Hill, Charles, and Marguerite Helmers. *Defining Visual Rhetorics*. Mahwah, NJ: Lawrence Erlbaum, 2004.

Hillers, John K. *"Photographed All the Best Scenery": Jack Hillers's Diary of the Powell Expeditions, 1871–1875*. Edited by Don D. Fowler. Salt Lake City: University of Utah Press, 1972.

Hoskins, Andrew. "Television and the Collapse of Memory." *Time & Society*, 13, no. 1 (2004): 109–27.

Huntford, Roland. *Shackleton*. New York: Carroll & Graf, 1998.

Jones, Max. *The Last Great Quest: Captain Scott's Antarctic Sacrifice*. Oxford: Oxford University Press, 2003.

Jurovics, Toby, ed. *Framing the West: The Survey Photographs of Timothy H. O'Sullivan*. Washington, D.C.; New Haven: Library of Congress: Smithsonian American Art Museum; Yale University Press, 2010.

Jussim, Estelle, and Elizabeth Lindquist-Cock. *Landscape as Photograph*. New Haven: Yale University Press, 1985.

Kammen, Michael G. *Mystic Chords of Memory: The Transformation of Tradition in American Culture*. New York: Knopf, 1991.

———. *Visual Shock: A History of Art Controversies in American Culture*. New York: Knopf, 2006.

Kelsey, Robin Earle. *Archive Style: Photographs & Illustrations for U.S. Surveys, 1850–1890*. Berkeley: University of California Press, 2007.

Keltner, Kathy. "From Myth to Metaphor to Memory: A Rhetorical Analysis of Televised Representations of Project Apollo, 1968–2004." Dissertation, Ohio University, 2007.

Kember, Sarah. "The Shadow of the Object: Photography and Realism." In *The Photography Reader*, edited by Liz Wells, 202–15. London: Routledge, 2003.

Kessler, Elizabeth. *Picturing the Cosmos: Hubble Space Telescope Images and the Astronomical Sublime*. Minneapolis: University of Minnesota Press, 2012.

King, Peter, ed. *Scott's Last Journey*. New York: HarperCollins, 2000.

Kloc, Joe. "The Case of the Missing Moon Rocks." *Atavist*, vol. 12 (February 2012).

Kluger, Jeffrey. *Apollo 8: The Thrilling Story of the First Mission to the Moon*. New York: Henry Holt and Company, 2017.

Kracauer, Siegfried. "Photography." Translated by Thomas Y. Levin. *Critical Inquiry* 19, no. 3 (Spring 1993): 421–36.

Kraft, Christopher C. *Flight: My Life in Mission Control*. New York: Dutton, 2001.

Krauss, Rosalind. "Photography's Discursive Spaces: Landscape/View." *Art Journal* 42, no. 4 (December 1, 1982): 311–19.

Kroes, Rob. *Photographic Memories: Private Pictures, Public Images, and American History*. Hanover, NH: Dartmouth College Press, 2007.

Kuehnel, Helmut A. *Apollo Experience Report—Photographic Equipment and Operations During Manned Space-Flight Programs*. Washington, D.C.: NASA, 1972.

Landsberg, Alison. *Prosthetic Memory: The Transformation of American Remembrance in the Age of Mass Culture*. New York: Columbia University Press, 2004.

Lansing, Alfred. *Endurance*. New York: Carroll & Graf, 2000.

Larson, Edward J. *An Empire of Ice: Scott, Shackleton, and the Heroic Age of Antarctic Science*. New Haven: Yale University Press, 2011.

Launius, Roger D. *Apollo: A Retrospective Analysis*. CreateSpace Independent Publishing Platform, 2012.

———. "Interpreting the Moon Landings: Project Apollo and the Historians." *History and Technology*, 22, no. 3 (2006): 225–55.

———. "Perceptions of Apollo: Myth, Nostalgia, Memory or All of the Above?" *Space Policy* 21, no. 2 (May 2005): 129–39.

———. "Public Opinion Polls and Perceptions of US Human Spaceflight." *Space Policy* 19, no. 3 (August 2003): 163–75.

———. "Why Go to the Moon? The Many Faces of Lunar Policy." *Acta Astronautica*, 70 (January 2012): 165–75.

Lazier, Benjamin. "Earthrise; or, The Globalization of the World Picture." *The American Historical Review* 116, no. 3 (June 1, 2011): 602–30.

Lazzarini, Alain. *Hasselblad and the Moon*. Granville, France: Self-Published, 2011.

Lembcke, Jerry. *The Spitting Image: Myth, Memory, and the Legacy of Vietnam*. New York: New York University Press, 1998.

Limerick, Patricia Nelson. *The Legacy of Conquest: The Unbroken Past of the American West*. New York: Norton, 1988.

Lindbergh, Anne Morrow. *Earth Shine*. New York: Harcourt, Brace & World, 1969.

Lindsay, John. *Lunar Stratigraphy and Sedimentology*. Amsterdam: Elsevier Scientific Publishing Group, 1976.

Lipsitz, George. *Time Passages: Collective Memory and American Popular Culture*. Minneapolis: University of Minnesota Press, 1990.

Logsdon, John M. *After Apollo? Richard Nixon and the American Space Program*. New York: Palgrave Macmillan, 2015.

———. *John F. Kennedy and the Race to the Moon*. New York: Palgrave Macmillan, 2010.

Lowenthal, David. "Past Time, Present Place: Landscape and Memory." *Geographical Review* 65, no. 1 (January 1975): 1–36.

Lowman, Paul. "Geologic Applications of Orbital Photography." NASA Technical Note D-4155. Washington, D.C.: NASA, April 1967.

———. "Geologic Orbital Photography: Experience from the Gemini Program." *Photogrammetria* 24, no. 3–4 (June 1969): 77–108.

Lubin, David M. *Shooting Kennedy: JFK and the Culture of Images*. Berkeley: University of California Press, 2003.

Mack, Pamela. *Viewing the Earth: The Social Construction of the Landsat Satellite System*. Cambridge, MA: MIT Press, 1990.

MacKenna, John. *Shackleton: An Irishman in Antarctica*. Madison: University of Wisconsin Press, 2002.

Maher, Neil. *Apollo in the Age of Aquarius*. Cambridge, MA: Harvard University Press, 2017.

———. "Shooting the Moon." *Environmental History* 9, no. 3 (July 2004): 526–32.

Mailer, Norman. *Of a Fire on the Moon*. Boston: Little, Brown, 1970.

Malina, Frank J. "On the Visual Fine Arts in the Space Age," *Leonardo* 3, no. 3 (July 1970): 323–25.

Marx, Leo. *The Machine in the Garden: Technology and the Pastoral Ideal in America*. Oxford: Oxford University Press, 2000.

Matusow, Allen J. *The Unraveling of America: A History of Liberalism in the 1960s*. New York: Harper & Row, 1984.

McCurdy, Howard E. *Space and the American Imagination*. Washington, D.C.: Smithsonian Institution Press, 1997.

McDougall, Walter A. . . . *the Heavens and the Earth: A Political History of the Space Age*. Baltimore: Johns Hopkins University Press, 1985.

McGregor, Alasdair. *Frank Hurley: A Photographer's Life*. New York: Viking, 2004.
McQuaid, Kim. "Selling the Space Age: NASA and Earth's Environment, 1958–1990." *Environment & History* 12, no. 2 (May 2006): 127–63.
McWilliams, John C. *The 1960s Cultural Revolution*. Westport, CT: Greenwood Press, 2000.
Middleton, David, and Derek Edwards, eds. *Collective Remembering*. London: Sage Publications, 1990.
Miller, Angela L. *The Empire of the Eye: Landscape Representation and American Cultural Politics, 1825-1875*. Ithaca: Cornell University Press, 1993.
Mindell, David A. *Digital Apollo: Human and Machine in Spaceflight*. Cambridge, MA: MIT Press, 2011.
Mirzoeff, Nicholas, ed. *The Visual Culture Reader*. London: Routledge, 2002.
Mitchell, William J. *The Reconfigured Eye: Visual Truth in the Post-Photographic Era*. Cambridge: MIT Press, 1994.
Mitchell, W. J. T., ed. *Landscape and Power*. Chicago: University of Chicago Press, 2002.
———. *Picture Theory: Essays on Verbal and Visual Representation*. Chicago: University of Chicago Press, 1994.
Moxey, Keith. "Visual Studies and the Iconic Turn." *Journal of Visual Culture* 7, no. 2 (2008): 131–46.
Naef, Weston J. *Era of Exploration: The Rise of Landscape Photography in the American West, 1860-1885*. Buffalo; Boston: Albright-Knox Art Gallery; distributed by New York Graphic Society, 1975.
National Aeronautics and Space Administration. *Earth Photographs from Gemini III, IV, and V*, NASA SP-129. Washington, D.C.: NASA, 1967.
———. *The Triumph of Astronaut L. Gordon Cooper, Jr. and the Faith 7*. Report #NASA-TM-85532. Houston: NASA, 1963.
Neal, Arthur G. *National Trauma and Collective Memory: Major Events in the American Century*. Armonk, NY: M. E. Sharpe, 1998.
Neal, Valerie. "Bringing Spaceflight Down to Earth: Astronauts and The IMAX Experience®." In *Spacefarers: Images of Astronauts and Cosmonauts in the Heroic Era of Spaceflight*, edited by Michael J. Neufeld, 149–74. Washington, D.C.: Smithsonian Institution Scholarly Press, 2013.
———. *Spaceflight in the Shuttle Era and Beyond: Redefining Humanity's Purpose in Space*. New Haven: Yale University Press, 2017.
Nelkin, Dorothy. *Selling Science: How the Press Covers Science and Technology*. New York: W. H. Freeman, 1995.
Neufeld, Michael J. *Spacefarers: Images of Astronauts and Cosmonauts in the Heroic Era of Spaceflight*. Washington, D.C.: Smithsonian Institution Scholarly Press, 2013.
Nora, Pierre. "Between Memory and History: Les Lieux de Mémoire." *Representations*, no. 26 (Spring 1989): 7–24.
Norris, Pippa. *Politics and the Press: The News Media and Their Influences*. Boulder, CO: Lynne Rienner Publishers, 1997.
Nugent, Robert, and Lowell Starr. "Gemini Photography Evaluation." NASA Technical Letter–69, March 1967.
Nye, David E. *American Technological Sublime*. Cambridge, MA: MIT Press, 1994.
O'Dell, C. R., Donald York, and Karl Henize, "Structure of the Barnard Loop Nebula as Determined from Gemini 11 Photographs." *The Astrophysical Journal* 150 (December 1967).

Orloff, Richard. *Apollo by the Numbers: A Statistical Reference*. Washington, D.C.: NASA, 2000.

Orvell, Miles. *American Photography*. Oxford: Oxford University Press, 2003.

Paltridge, George H. *Photographic Instruction Text: A Systematic Course and Working Guide in All the Processes Which Ordinarily Take Up the Attention of Camera Workers*. New York: Photo-Text Press, 1900.

Pearlman, Robert. "Sigma 7 at 50: Retro Space Images Recall 5th U.S. Spaceflight." *Space .com*, October 3, 2012.

Perez, Renato. "Prestige is High in Space Photos." *New York Times*, September 25, 1966.

Pieters, C., and T. B. McCord. "Characterization of Lunar Mare Basalt Types. A Remote Sensing Study Using Reflection Spectroscopy of Surface Soils." In *Lunar Science Conference, 7th, Houston, Tex., March 15-19, 1976, Proceedings*. Volume 3. New York: Pergamon Press, 1976.

Poole, Robert K. *Earthrise: How Man First Saw the Earth*. New Haven: Yale University Press, 2008.

Preston, Diana. *A First Rate Tragedy: Robert Falcon Scott and the Race to the South Pole*. Boston: Houghton Mifflin, 1998.

Prown, Jules David, et al. *Discovered Lands, Invented Pasts: Transforming Visions of the American West*. New Haven: Yale University Press, 1992.

Pyne, Stephen J. *The Ice: A Journey to Antarctica*. Iowa City: University of Iowa Press, 1986.

Quartermain, L. B. *South to the Pole: The Early History of the Ross Sea Sector, Antarctica*. Oxford: Oxford University Press, 1967.

Reidy, Michael S., Gary Kroll, and Erik M. Conway. *Exploration and Science: Social Impact and Interaction*. Santa Barbara, CA: ABC-CLIO, 2007.

Respini, Eva. *Into the Sunset: Photography's Image of the American West*. New York: Museum of Modern Art, 2009.

Robinson, Michael. *The Coldest Crucible: Arctic Exploration and American Culture*. Chicago: University of Chicago Press, 2006.

Ryan, James R. *Picturing Empire: Photography and the Visualization of the British Empire*. Chicago: University of Chicago Press, 1997.

Sandweiss, Martha A. *Print the Legend: Photography and the American West*. New Haven: Yale University Press, 2002.

Savours, Ann, ed. *Scott's Last Voyage, Through the Antarctic Camera of Herbert Ponting*. New York: Praeger Publishers, 1975.

Schick, Ron, and Julia Van Haaften. *The View from Space: American Astronaut Photography, 1962–1972*. New York: C. N. Potter, distributed by Crown, 1988.

Schirra, Jr., Walter M., and Richard N. Billings, *Schirra's Space*. Boston: Quinlan Press, 1988.

Scott, David Meerman, and Richard Jurek. *Marketing the Moon: The Selling of the Apollo Lunar Program*. Cambridge, MA: MIT Press, 2014.

Scranton, Philip. "Behind the Icon: NASA's Mercury Capsules as Artefact, Process, and Practice." In *Showcasing Space*, Vol. 6, edited by Martin J. Collins and Douglas Millard, 50–75. London: Science Museum, 2005.

———. "None-Too-Porous Boundaries: Labor History and the History of Technology." *Technology and Culture* 29, no. 4 (October 1988).

Shales, Tom. "'Big Blue Marble': Picking Up a 'Blue Marble'." *Washington Post,* May 3, 1974.

Simpson-Housley, Paul. *Antarctica: Exploration, Perception, and Metaphor*. New York: Routledge, 1992.

Smith, Michael L. "Selling the Moon: The U.S. Manned Space Program and the Triumph of Commodity Scientism." In *The Culture of Consumption: Critical Essays in American History, 1880–1980*, edited by Richard Wightman Fox and T. J. Jackson Lears, 177–209. New York: Pantheon Books, 1983.

Solomon, Susan. *The Coldest March: Scott's Fatal Antarctic Expedition*. New Haven: Yale University Press, 2001.

Sonnevend, Julia. "Iconic Rituals: Towards a Social Theory of Encountering Images." In *Iconic Power: Materiality and Meaning in Social Life*, edited by J. C. Alexander, et al., 219–32. New York: Palgrave Macmillan, 2012.

Sontag, Susan. *On Photography*. New York: Farrar, Straus and Giroux, 1977.

Stafford, Thomas, and Michael Cassutt, *We Have Capture: Tom Stafford and the Space Race*. Washington, D.C.: Smithsonian Institution Press, 2002.

Stegner, Wallace. *Beyond the Hundredth Meridian: John Wesley Powell and the Second Opening of the West*. New York: Penguin Books, 1992.

Steven-Boniecki, Dwight. *Live TV from the Moon*. Burlington, Ontario: Apogee Books, 2010.

Sturken, Marita. *Practices of Looking: An Introduction to Visual Culture*. Oxford: Oxford University Press, 2001.

———. *Tangled Memories: The Vietnam War, the AIDS Epidemic, and the Politics of Remembering*. Berkeley, CA: University of California Press, 1997.

Swenson, Loyd S., Charles C. Alexander, and James M. Grimwood. *This New Ocean: A History of Project Mercury*. NASA SP 4201. Washington, D.C.: National Aeronautics and Space Administration, NASA History Office, Office of Policy and Plans, 1998.

Tagg, John. "Evidence, Truth, and Order: A Means of Surveillance." In *Visual Culture: The Reader*, edited by Jessica Evans and Stuart Hall, 244–73. London: Sage/Open University, 1999.

Tannenbaum, Rob, and Craig Marks. *I Want My MTV: The Uncensored Story of the Music Video Revolution*. New York: Penguin, 2011.

Thomson, John. "Photography and Exploration." *Proceedings of the Royal Geographical Society and Monthly Record of Geography*, New Monthly Series, 13, no. 11 (November 1891): 669–75.

Trachtenberg, Alan. *Reading American Photographs: Images as History, Mathew Brady to Walker Evans*. 1st ed. New York: Hill and Wang, 1989.

Truettner, William H., and Nancy K. Anderson, eds. *The West as America: Reinterpreting Images of the Frontier, 1820–1920*. Washington, D.C.: Smithsonian Institution Press, 1991.

Turner, Fred. *From Counterculture to Cyberculture: Stewart Brand, the Whole Earth Network, and the Rise of Digital Utopianism*. Chicago: University of Chicago Press, 2008.

"The Two New Hasselblads." *Modern Photography*, December 1954.

Underwood, Richard. "Space Photography." In *Gemini Summary Conference, February 1–2, 1967*. Houston: NASA, 1967.

van der Meer Mohr, H. E. C. "Geological Interpretation of Hyperaltitude Photographs from Gemini Spacecraft." *Photogrammetria* 24, no. 3–4 (June 1969): 167–74.

van Dijck, José. "Mediated Memories: Personal Cultural Memory as Object of Cultural Analysis." *Continuum: Journal of Media & Cultural Studies* 18, no. 2 (2004): 261–77.

Van Riper, Frank. "Space Cameras: From Ansco to Hassy to Digital." *Washington Post*, October 23, 2003.

Watkins, Billy. *Apollo Moon Missions: The Unsung Heroes*. Westport, CT: Praeger Publishers, 2006.

Wells, Liz. *Land Matters: Landscape Photography, Culture and Identity*. London: I. B. Tauris, 2011.

Wheeler, George M. *Wheeler's Photographic Survey of the American West, 1871–1873*. New York: Dover Publications, 1983.

White, Richard. *"It's Your Misfortune and None of My Own": A History of the American West*. Norman: University of Oklahoma Press, 1991.

Wilkinson, Alec. *The Ice Balloon: S. A. Andree and the Heroic Age of Arctic Exploration*. New York: Knopf Doubleday Publishing Group, 2013.

Wobber, Frank J. "Space Photography: A New Analytical Tool for the Sedimentologist." *Sedimentology* 9 (1967).

Worden, Al. *Falling to Earth: An Apollo 15 Astronaut's Journey*. Washington, D.C.: Smithsonian Books, 2011.

Worster, Donald. *A River Running West: The Life of John Wesley Powell*. Oxford: Oxford University Press, 2001.

Wrobel, David. *The End of American Exceptionalism: Frontier Anxiety from the Old West to the New Deal*. Tucson: University of Arizona Press, 1993.

Wyckoff, C. W., and J. C. McCue. *A Study to Determine the Optimum Design of a Photographic Film for the Lunar Surface Hand-Held Camera*. Prepared for NASA by EG&G, Inc., June 2, 1965.

Index

Illustrations and figures are indicated by page numbers in italics. Tables are indicated by "t" after the page number.

About the Author

Jennifer K. Levasseur is a museum curator in the Department of Space History at the Smithsonian National Air and Space Museum in Washington, DC. She received her BA in history from the University of Michigan in 1999, an MA in American studies from George Washington University in 2002, and a PhD in history from George Mason University in 2014. During nearly two decades with the museum, Levasseur has worked on artifact loans and digital and exhibition projects. She serves as program committee chair of the Mutual Concerns of Air and Space Museums Conference, a biennual gathering of staff from museums around the world. Most recently, she curated the exhibit *Outside the Spacecraft: 50 Years of Extra-Vehicular Activity* and is the exhibition curator for *Moving Beyond Earth*, the museum's exhibit on the Space Shuttle, International Space Station, and future of human spaceflight. Her collections responsibilities include astronaut cameras, chronographs, the Space Shuttle and International Space Station program collections. She is married to Ben Levasseur, who works for the U.S. Navy in foreign military sales, and is mother to Remi and Drew, two aspiring hockey players.